# The Magic Furnace

Also by Marcus Chown

**AFTERGLOW OF CREATION**

# THE MAGIC FURNACE

## The Search for the Origins
## of Atoms

Marcus Chown

OXFORD

UNIVERSITY PRESS

2001

# OXFORD
## UNIVERSITY PRESS

Oxford New York
Athens Auckland Bangkok Bogotá Buenos Aires Calcutta
Cape Town Chennai Dar es Salaam Delhi Florence Hong Kong Istanbul
Karachi Kuala Lumpur Madrid Melbourne Mexico City Mumbai
Nairobi Paris São Paulo Shanghai Singapore Taipei Tokyo Warsaw Toronto

and associated companies in
Berlin Ibadan

Published by Oxford University Press, Inc.
198 Madison Avenue, New York, New York 10016

First published in Great Britain in 1999 by Jonathan Cape

Oxford is a registered trademark of Oxford University Press

Library of Congress Cataloging-in-Publication Data is available
ISBN 0-19-514305-1

2 4 6 8 9 7 5 3 1
Printed in the United States of America
on acid-free paper

# Contents

To my dad and mum, who never doubted I would be Albert Einstein, win the Nobel prize and punch out a bestseller.

# Acknowledgements

My thanks to the following people who either helped me directly, inspired me, or simply encouraged me during the writing of this book. Karen, my mum and dad, Neil Belton, Charlotte Mendelson, Murray Pollinger, Sara Menguc, Sir Fred Hoyle, Harald Fritzch, the late Roger Tayler, Robert Wilson, David Tytler, Jordi Miralda-Escude, Peter Kalmus, John Mather, Sir Martin Rees, Max Tegmark, John Emsley, Ken Croswell, Juliet Walker, Nigel Henbest, Nick Booth, Anne Bean, Peter Fink, Elisabeth Geake, Nick Mayhew-Smith, Michael White, Stephen Hedges, David Hough, Pauline and David Parslow, Pat and Brian Chilver. It goes without saying, I hope, that none of these people are responsible for any errors.

# *Prologue*

## THE COSMIC CONNECTION

I believe a leaf of grass is no less than the journeywork of the stars.

Walt Whitman

Every breath you take contains atoms forged in the blistering furnaces deep inside stars. Every flower you pick contains atoms blasted into space by stellar explosions that blazed brighter than a billion suns. Every book you read contains atoms blown across unimaginable gulfs of space and time by the wind between the stars.

Astronomers often talk glibly of black holes and exploding stars, pulsars, quasars and the titanic eruption of the big bang. But if the truth be told it is extremely difficult to believe that any of these things are actually real – as real, for instance, as a mountain or an oak tree or a newborn baby. They are simply too remote, too far removed from the familiar world of our experience. It seems inconceivable that they could have the slightest connection with our everyday lives.

But this is an illusion.

Many of the most dramatic and awe-inspiring of cosmic events – from the violent death throes of stars to the titanic fireball that gave birth to the entire universe 15 billion years ago – are connected to us *directly* by way of the atoms that make up our bodies.

If the atoms that make up the world around us could tell their stories, each and every one of them would sing a tale to dwarf the greatest epics of literature. From carbon, baked in bloated red giants – stars so enormous they could swallow a million suns – to uranium, cooked in supernova explosions – just about the most violent cataclysms in all of creation. From boron, generated in atom-crunching collisions in the deep-freeze of interstellar space, to helium, forged in the hellish first few minutes of the big bang itself.

The iron in your blood, the calcium in your bones, the oxygen that fills your lungs each time you take a breath – all were baked in the fiery ovens deep within stars and blown into space when those stars grew old, and perished. Every one of us is a memorial to long-dead stars. Every one of us was quite literally made in heaven.

For thousands of years, astrologers have been telling us that our lives are controlled by the stars. Well, they were right in spirit if not in detail. For science in the twentieth century has revealed that we are far more intimately connected to events in the cosmos than anyone ever dared imagine. Each and every one of us is stardust made flesh.

The story of how we discovered the astonishing truth of our cosmic origins – how we found the magic furnace that forged the atoms – is one of the great untold stories of science. In fact, it is two stories intertwined: the story of atoms and the story of stars. Neither story can be told without the other. For the stars contain the key to unlocking the secret of atoms and the atoms the solution to the puzzle of stars.

The story of the quest for the origin of atoms is the story of two great theories, and the pendulum that has swung back and forth between them. One theory maintained that atoms were cooked inside stars then ejected into space to provide the raw material for new suns and new planets, while the other theory contended that atoms were assembled at the very birth of the universe, in the first blisteringly hot minutes of the big bang.

At first the pendulum swung to stars as the most likely site of the elusive magic furnace. Then, when it appeared that stars were simply not hot enough for the job of cooking atoms, the pendulum swung to the big bang. When the big bang turned out not to be up to the job either, the pendulum swung back to stars again – or at least most of the way to stars. For nature, as we are so often reminded, is under no obligation to make things simple just for our convenience.

But before we were in any position to discover the cosmic origin of atoms, we first needed to realise that atoms were actually made and not put in the universe on Day One by the Creator. And before we could realise this truth we needed to realise something even more basic and far from obvious: that everything is made of atoms . . .

# Part One
# Atoms

# I

# The Alphabet of Nature

## HOW WE DISCOVERED THAT EVERYTHING ON EARTH IS MADE OF ATOMS.

If in some cataclysm, all of scientific knowledge were to be destroyed, and only one sentence passed on to the next generations of creatures, what statement would contain the most information in the fewest words? Everything is made of atoms.

Richard Feynman

> To see a world in a grain of sand
> And a heaven in a wild flower,
> Hold infinity in the palm of your hand
> And eternity in an hour.

William Blake

In the mists of antiquity, it must have occurred to many people to ask the question: what happens if I take this stick, this piece of cloth, this clay tablet, and cut it in half, then in half again? Can I go on forever? Or will I eventually reach a point when I will be unable to cut it any smaller? The first person to record an answer to this question was the Greek philosopher Democritus.

Little is known about Democritus. His ideas have come down to us solely through the writings of others. He was born around 470 BC, but no one knows quite where. He travelled extensively throughout the Mediterranean and founded a school at Abdera in Thrace. Nowadays, the site is occupied by the Greek town of Avdhira near the border with Bulgaria, but in the fifth century BC it was a prosperous and bustling port on the shores of the Aegean.

Democritus was obsessed by a single question: what is the nature of matter? The question had first been posed more than a century earlier by Thales of Miletus, the founding father of Greek philosophy, but its imprecise wording had prevented a satisfactory answer. Democritus' genius was to refine Thales' question. In the course of relentless, exhausting discussions with his teacher, Leucippus, he transformed what was a vague enquiry into a question of exquisite precision for which there could be only two alternative answers: Could matter be subdivided forever?

Democritus' answer was an emphatic *no*. It was absolutely inconceivable to him that any material object could be cut into smaller and smaller parts without limit. Sooner or later, he reasoned, such a cleaving process must result in a grain of matter that could not be made any smaller. Since the Greek phrase for 'uncuttable' was *a-tomos*, Democritus christened the indestructible grains out of which everything was made 'atoms'. 'By convention there is sweet, by convention there is bitterness, by convention hot and cold, by convention colour,' he wrote. 'But in reality there are only atoms and the void.'

Atoms and the void.

The idea instantly made sense of several features of the world that were inexplicable if matter were continuous rather than grainy. Where, for instance, did salt go when it was stirred into a pot of warm water? If Democritus was right, it simply disintegrated into its constituent atoms which then lodged themselves in the empty spaces between the atoms of water. The atomic idea also explained how it was possible for fish to swim through the sea. If water was a continuous material, there would be no gaps for a fish to slip through. However, if the world was made of atoms separated by empty space, the tip of a fish's nose could slip between the atoms of the water, parting the liquid like a curtain as the fish swam forward.

There was no doubt that atoms could explain some puzzling phenomena. But in truth they were merely one man's daydreams. Atoms, if they really existed, were far too small to be perceived directly by the senses. How then would it ever be possible to establish their reality? Fortunately, there was a way.

The trick was to assume that atoms existed, then deduce a logical consequence of this assumption for the everyday world. If the consequence matched reality, then the idea of atoms was given a boost. If it did not, then it was time to look for a better idea.

In fact, this was precisely the same kind of argument that Democritus had used to support his revolutionary idea. By first assuming that atoms existed, he had deduced that salt should dissolve in water and that fish should have no difficulty swimming through the sea, two observations that both accorded with reality.

However, the ability of different substances to penetrate each other was merely a 'quality' of matter. If the idea of atoms were to be put on a firmer footing it would be necessary to deduce from the existence of atoms some measurable property of matter – a 'quantity' that could be gauged with a ruler or a set of scales or some other kind of measuring instrument. However, deducing a precise property of matter was impossible without a precise picture of atoms.

Democritus had envisaged free atoms as flying about ceaselessly through empty space. What was needed therefore was a precise picture of how atoms flew about through space. Such a picture required a knowledge of the laws that governed all motion. However, their formulation was well beyond the capabilities of Democritus. It would have to await the rise of science.

## THE RISE OF SCIENCE

The Greeks, for all their dazzling brilliance, did not invent science. Although thinkers such as Democritus speculated endlessly about the great overarching principles that controlled the cosmos, they were fatally handicapped by a reluctance to test their speculations by prodding and probing the world around them. At the other end of the spectrum were the craftsmen, who fired and glazed pottery, who forged weapons out of bronze and iron. They prodded and probed the world but were handicapped by their reluctance to speculate about the principles that governed their craft.

For thousands of years the two traditions remained quite separate – the scholarly tradition, epitomised by those who theorised about the world but avoided getting their hands dirty, and the craft tradition, typified by those who obtained their knowledge of the world from solid hands-on experience but who developed no theory of what they were doing that might indicate better ways of doing it.

The two traditions were like two great rivers rolling down to the sea. Their eventual convergence might have been expected to create a bigger, more powerful river. What it in fact created was an overwhelming, unstoppable flood.

The third tradition, which emerged from the union of the scholarly and craft traditions, was of course science.

Science was an immensely powerful method of investigating the world. It involved carrying out careful experiments, with nature itself as the ultimate arbiter of which theories were right and which were wrong. In the sixteenth and seventeenth centuries, the revolutionary new method began to be practised by a small band of far-sighted men, few of whom at first recognised the power of the tool they were wielding. Of this small band, none used the tool to greater effect than Isaac Newton.

## THE LAWS OF MOTION

Just as all roads once led to Rome, all paths in science invariably lead back to Newton, arguably the most powerful intellect the world has known. 'Nature was to him an open book, whose letters he could read without effort,' wrote Albert Einstein. Newton viewed the universe as a giant riddle set by the Creator which might be cracked by the relentless application of pure thought. In this task he was aided by almost superhuman powers of concentration, which enabled him to hold an abstract problem in the forefront of his mind for weeks until it finally yielded its secrets. Like the Greek philosophers before him, Newton attempted to discern the universal principles that governed the world. Unlike Democritus and his contemporaries, however, he carried out experiments both to test his theories and to discover new phenomena.

Newton's investigations into the way bodies moved led him to his famous 'laws of motion', which govern how massive objects respond when subjected to forces. By applying the laws to the everyday world, Newton was able to explain the fall of weights dropped from tall towers, the flight of cannon balls shot through the air and the recoil of bowling balls involved in head-on collisions.

But Newton did not stop here. He took his laws of motion and carried them to an entirely different domain: the domain of the solar system, where very large bodies such as the moon and planets moved under the influence of the invisible but all-pervasive force of gravity. In doing so, Newton was able to explain why the moon raises tides twice a day in the world's oceans, and why the planets trace out elliptical paths around the sun.

In fact, so successful were Newton's laws of motion that they became a standard part of the tool-kit of science. One man who used them to brilliant effect was the eighteenth-century Swiss mathematician Daniel Bernoulli. If Newton's genius was to take his laws of motion and apply them to the domain of the very large, Bernoulli's was to apply them to the domain of the very small: to the world of atoms.

# ATOMS IN MOTION

Bernoulli is most famous for his discovery that the 'pressure' of a liquid or gas drops when it is forced to flow rapidly. The 'Bernoulli effect' is exploited every day by aircraft whose wings are shaped so that air flows faster over their top surfaces than their bottom surfaces. It is the excess of pressure pushing upwards on a wing that provides the 'lift' necessary for heavier-than-air flight.

But Bernoulli did a lot more than simply pioneer the field of 'fluid flow'. He also carried out a ground-breaking investigation of atoms and their consequences for the measurable properties of matter. Bernoulli had not the slightest idea what atoms looked like, or even how big they were. However, he had one big advantage over Democritus. He knew that free atoms were more than simply tiny grains flying about through space; they were tiny grains flying about through space and obeying Newton's laws of motions. To make the most of this insight, Bernoulli needed to find a situation in nature where a precise knowledge of the way in which atoms flew about might lead to a prediction of a measurable property of matter. He identified one in the case of a gas, which he visualised as a host of tiny grains in perpetual frenzied motion like a swarm of angry bees.

Bernoulli's brainwave was to realise that the atoms of such a gas would hammer relentlessly on the walls of any containing vessel. The effect of each individual impact would of course be vanishingly small. However, the effect of billions upon billions of atoms hammering away incessantly would be to push the walls back. A gas made of atoms would therefore exert a jittery force which our coarse senses would feel as an average push or 'pressure'.

This was Bernoulli's great insight: to connect the small-scale behaviour of the atoms of a gas to its large-scale pressure.

The pressure of a gas such as steam was easy to measure. It was necessary only to introduce the steam into a hollow cylinder containing a 'piston'. In essence this was a movable wall which could travel up and down the cylinder in response to the pressure of the steam. The farther the piston moved along the cylinder, the higher was the steam pressure. The pressure of a gas provided a direct link between the world of human experience – where pistons could actually be seen to move – and the invisible world of atoms. But to make the link explicit, Bernoulli needed to use his picture of drumming atoms to deduce how the pressure of a gas should behave in different circumstances – for instance, if the gas was compressed or heated.

Bernoulli first made some simplifying assumptions. For instance, he assumed that atoms were very small compared with the gulf between them. This turns out to be a very good assumption. On average the atoms in the air around us are separated by several hundred times their diameters. If the atoms of a gas were blown up to the size of tennis balls, for instance, there would be barely 100 flying about in the volume of a large hall. The assumption that the atoms in a gas are a long way apart enabled Bernoulli to ignore any force that existed between them; any such a force – whether of attraction or repulsion – was unlikely to be long-range.* With the motion of each atom unaffected by its fellows, Newton's laws ordained that it should fly at a constant speed in a straight line. The exception to this was of course when it slammed into a piston or the walls of a container. Bernoulli assumed that in such a collision a gas atom simply bounced off the surface without losing any speed whatsoever, in the process imparting a minuscule force to the wall.

Bernoulli now asked himself: what would happen to the pressure of a gas in a cylinder if someone squeezed the gas by pushing in the piston? To answer the question, he imagined pushing such a piston until the gas was compressed into half its original volume. Since the gas atoms would now have to fly only half as far as before between collisions, in any given interval of time they would collide with the piston twice as many times. Consequently, they would exert double the pressure. Similarly, if the gas were compressed to a third of its volume, its pressure would triple. This was exactly the way a real gas behaved. It had been observed by the English scientist Robert Boyle in 1660, and been named Boyle's law in his honour.

Bernoulli next asked: what would happen to the pressure of a gas in a cylinder if the gas was heated while its volume was left unchanged?

To answer this question, he exploited a remarkable insight which would not be generally accepted by the rest of the scientific community for more than a century. His insight was that the temperature of a gas was merely a measure of how fast on average its atoms were flying about. When a gas was heated, its atoms were simply speeded up.

Bernoulli imagined heating the steam in a cylinder with the piston in place. Since the atoms would now be moving faster, they would collide with the piston more often and with greater force. Consequently, the pressure of the gas would rise. This would be obvious to anyone trying to hold the piston in place because they would have to struggle harder

---

* The kind of gas Bernoulli imagined is often called an 'ideal' or 'perfect' gas. This is to distinguish it from a 'real' gas whose behaviour may in some circumstances differ from the ideal – for instance, when the gas is subjected to very high pressure. The gas particles may then be squeezed so close together that they are seriously affected by the forces between them. Any attempt to understand the gas must therefore take this into consideration.

to keep it from moving along the cylinder. Once again, this was precisely the way that a real gas behaved. It had been observed by the French scientist Jacques Alexandre César Charles in 1787, and christened Charles's law.

Bernoulli had triumphantly predicted two measurable properties of a gas – the way its pressure went up when its volume went down and the way its pressure went up when its temperature went up. And he had done it by simply assuming that a gas consisted of countless atoms which flew hither and thither and drummed on the walls of their containing vessel like hailstones on a tin roof. In the words of Piet Hein:

> Nature, it seems, is the popular name
> for millards and millards and millards
> of particles playing their infinite game
> of billards and billards and billards.

In the late nineteenth century, Bernoulli's method of deducing the properties of a gas from the collective behaviour of all of its atoms was taken to its logical conclusion by both James Clerk Maxwell in Britain, and Ludwig Boltzmann in Germany. But although Maxwell and Boltzmann's work provided the most convincing evidence yet that the world was composed of tiny grains of matter, the existence of atoms was far from generally accepted and remained the subject of intensely bitter debate well into the twentieth century.

Those who disputed the existence of atoms had strong convictions about what did and did not constitute science. The conviction, held most notably by the Austrian physicist Ernst Mach, was that science had no business to be concerning itself with any feature of the world which could not be observed directly with the senses.* Since nobody had actually *seen* an atom – nor were they ever likely to – Mach maintained that the whole atomic concept was unscientific and should be ruthlessly rooted out of science. When an army of scientific zealots, inspired by Mach's view, mounted a savage crusade against the proponents of atoms, it all proved too much for Boltzmann. Prone to bouts of depression and born with one skin too few, he succumbed to the pressure and committed suicide while on holiday in 1906.

Ironically, the final proof of the existence of atoms had come the year before Boltzmann took his life. It was provided by an obscure clerk in the Swiss patent office. His name was Albert Einstein.

---

* He is today remembered by all pilots of aircraft that fly faster than the speed of sound, otherwise known as Mach 1.

## THE CRAZY DANCE OF POLLEN GRAINS

The year 1905 was a miraculous year for the 25-year-old Einstein. In the space of twelve months, he published four trail-blazing papers. One was on the revolutionary new theory of 'special relativity' which redefined space and time; a second showed how to deduce the size of molecules from the behaviour of liquids; a third addressed the particle-like nature of light. The fourth paper has been a little overshadowed by the others but was nevertheless enormously significant. For it proved, once and for all, the reality of atoms. More specifically, it made sense of a baffling observation made almost a century earlier by a Scottish botanist called Robert Brown.

Brown, who had sailed to Australia on the Flinders expedition of 1801, had classified 4000 species of antipodean plants, in the process discovering the 'nucleus' of living cells. But his greatest discovery had come in 1827. While looking through a microscope at pollen grains floating in water, he was amazed to see that the grains were jiggling about as if something was repeatedly kicking them. The behaviour became known as 'Brownian motion'. He could think of no plausible explanation and nor could anyone else.

Einstein's genius was to realise that each pollen grain was indeed being kicked – by atoms or, more precisely, by molecules of water.[*] At a mere thousandth of a millimetre across, a pollen grain was small enough to be jostled by the very building blocks of matter. It was as if a giant inflatable rubber ball, taller than a person, was being pushed about a field by a large number of people. If each person was pushing in their own direction, without the slightest regard to their companions, at any instant there were likely to be slightly more people on one side than on the other. The imbalance would be enough to cause the ball to move erratically about the field. Similarly, the erratic motion of a pollen grain could be explained if at every moment there were slightly more water molecules bombarding it from one side than from another.

Einstein devised a mathematical theory to describe Brownian motion. Its predictions were triumphantly confirmed three years later by the French scientist Jean-Baptiste Perrin who, for convenience, replaced pollen grains with particles of gamboge, a yellow gum resin from a Cambodian tree.

Einstein's theory predicted how far and how fast the average pollen grain should travel in response to the relentless battering it was receiving from water molecules all around. Everything hinged on the size of the water molecules; the bigger they were, the bigger would be the

---

[*] The molecules each consist of two atoms of hydrogen attached to an atom of oxygen.

imbalance of forces on a pollen grain, and the more exaggerated its consequent Brownian motion.

By comparing his observations of gamboge particles through a microscope with the predictions of Einstein's theory, Perrin was able to deduce the size of water molecules, and hence of the atoms out of which they were built. His conclusion was that atoms were only about a 10 billionth of a metre across – so small that it would take 10 million, arranged end to end, to span the width of a full stop on this page. Einstein and Perrin had found the most direct evidence yet for the existence of atoms. No one who peered into a microscope and saw the crazy dance of pollen grains under relentless machine-gun bombardment could now doubt that the world was really made from tiny, bullet-like particles.

But Brownian motion revealed only the combined effect of large numbers of particles on bodies which were far larger than atoms. The fundamental building blocks of all matter remained stubbornly out of sight.

Atoms were a mere 10 millionth of a millimetre across. The possibility of seeing them directly might be entertained by science-fiction writers, but not by reputable scientists. Science fiction, however, has a peculiar habit of coming true. In 1980, two physicists in Switzerland invented and built one of the most remarkable instruments in the history of science. Using it, Gerd Binnig and Heinrich Rohrer became the first people in history to actually 'see' an atom.

## SEEING ATOMS

The instrument that fulfilled Democritus' 2000-year-old dream was called the 'scanning tunnelling microscope', or STM for short. It was born in the autumn of 1978 when Binnig, a 31-year-old German doctoral student, was putting the finishing touches to his thesis at Wolfgang Goethe University in Frankfurt.

Binnig was interested in the surfaces of 'semiconductor' materials such as silicon, which formed the foundations of computer chips. It was an interest which happened to be shared by Heinrich Rohrer, a middle-aged Swiss physicist who was visiting Binnig's university from IBM's research laboratory in Zurich. When the two men bumped into each other one day, their conversation turned to the prospects of ever being able to see the fine details of surfaces like silicon. Such a feat, if possible, would be a boon to computer manufacturers, who were constantly trying to shrink transistors and other electronic components and pack them closer together on the surface of chips. In this task, they were severely hampered by their ignorance of what such surfaces looked like on a very small scale. They were like gods who towered above the miniature landscape of their world but whose eyes were hopelessly blindfolded.

But even a blindfolded god has one means open to him to determine the lie of the land. He can use his sense of touch to feel the ups and downs of hills and valleys, and in this way build up a mental picture of the landscape. By running a giant finger over the ground, he might even be able to sense features as small as trees and buildings. Using a finger to explore the submicroscopic landscape of a material like silicon might seem a little fanciful. But, in essence, this was the idea that occurred to Binnig as he talked with Rohrer. Instead of a finger of flesh and blood, however, he envisioned a finger of metal – a very fine needle, like the stylus of an old-fashioned record player.

Of course, there was no way a needle could actually feel a surface like a human finger. However, if the needle were charged with electricity and placed extremely close to the surface of a metal or semiconductor, a minuscule, but measurable, electric current would leap the gap between the tip of the needle and the surface. It was known as a 'tunnelling current', and it had a crucial property which Binnig realised might be exploited: the current was extraordinarily sensitive to the width of the gap. If the needle were moved even a shade closer to the surface, the current would grow very rapidly; if it were pulled away a fraction, the current would plummet. The size of the tunnelling current therefore revealed the distance between the needle tip and the surface: it gave the needle an artificial sense of touch.

Rohrer was so impressed by Binnig's idea that he invited him to Zurich to transform it into reality. It was the start of an immensely productive partnership which would ultimately lead Binnig and Rohrer to Stockholm to receive the 1986 Nobel prize for physics.

The first problem was to find a needle fine enough to feel the submicroscopic details of a surface. A needle is insensitive to features much smaller than itself, just as a finger is insensitive to the fine grooves on an old-fashioned vinyl record. Common sense therefore implied that a needle tip which could feel the undulations of individual atoms would itself have to be only a few atoms across. Unfortunately, this was hundreds of times finer than the finest needle in existence. However, in 1979, Binnig made a remarkable discovery. He found that the tunnelling current leapt to a metal surface from only a tiny patch of atoms at the very tip of the needle. It meant that a needle was actually tens of times sharper than it appeared. In fact, when Binnig and Rohrer made needles out of the metal tungsten, they discovered that the tips invariably consisted of a protruding clump of only a few atoms. With such needles it would be possible to sense features smaller than either man had ever dared hope.

But turning Binnig's idea into reality required more than an ultra-fine needle. It required an elaborate scaffolding of springs and shock absorbers to hold the needle just a whisker above the surface of a material and isolate it from stray vibrations. On the scale of atoms, even the footsteps of someone in the same building or the passage of a car down a nearby street would seem like a major earthquake. To control the height of the needle, Binnig and Rohrer exploited the tunnelling current itself. They arranged that if the current fell the needle would be automatically lowered and if the current grew the needle would be pulled up. In this way, Binnig and Rohrer were able to keep their needle at a constant height as it tracked back and forth across the surface of a material.

It was as if lightning flickered from the finger of a god to the ground. If he lifted his finger too high, the lightning died away until he had no sense of the surface; if he moved it too close, the lightning grew to a painful intensity. By keeping the lightning crackling at a tolerable level, he was able to follow the ups and downs of the terrain with his finger.

A god had the option of building up a picture of the miniature landscape in his imagination. However, no such possibility was open to Binnig and Rohrer. Instead, the two physicists had to resort to a computer to convert the up-and-down motion of their needle into a visual image. When they did so, what they saw on their computer screen in Zurich took their breath away.

It was one of the most remarkable images in the history of science. It was an image to rank with the image of the earth rising above the grey desolation of the moon, or with the sweeping spiral staircase of DNA. For it was the first ever picture of the invisible realm that underpinned the everyday world. Here, at long last, were atoms in all their microscopic glory.* They looked like tiny footballs. They looked like oranges, stacked in boxes row on row. But, most of all, they looked like the tiny hard grains of matter which one man had seen so clearly in his mind's eye two and a half thousand years earlier.

---

* It should be pointed out that there is some ambiguity in STM pictures because the tunnelling current depends not only on the hills and valleys of a surface, but also on its electrical charge. In the beginning, this caused some critics to question what the STM was actually seeing. Fortunately, the STM was merely the first of a whole armoury of 'super microscopes'. When used together, they leave scientists in little doubt that they are really seeing atoms.

The scanning tunnelling microscope revealed Democritus' tiny motes of matter, whose graininess explained where salt went when it dissolved in water and how fish swam through the sea. It revealed Bernoulli's hard little balls, whose relentless hammering on a piston made sense of the behaviour of a gas. It revealed Einstein's tiny bullet-like particles, whose machine-gun bombardment of pollen grains explained the frenetic dance of Brownian motion. But, for all its spectacular success, the scanning tunnelling microscope revealed only one side of atoms. As Democritus himself had realised, atoms were a lot more than simply grains in motion.

## THE ALPHABET OF NATURE

Democritus had imagined atoms coming in a number of different kinds – which differed in their size and shape and perhaps their weight. By arranging these various types in different patterns, it was possible to make a rose, a bar of gold or a human being. Atoms, in short, were the alphabet of nature. If Democritus was right, the bewildering complexity of the world was nothing more than an elaborate illusion. It was merely a consequence of the myriad ways in which a handful of fundamental building blocks of matter could be put together.

It was one of the most breathtaking leaps of the imagination in history. With the power of thought alone, Democritus had lifted a corner of the veil that shrouded the world from our senses. He had found that, underneath it, reality was remarkably simple.

The key step in proving such a revolutionary idea would of course be identifying the different kinds of atom. However, the fact that atoms were far too small to be perceived directly by the senses made the task every bit as formidable as proving that atoms were tiny grains of matter in ceaseless motion. In the circumstances, the only possibility was to find substances that were made exclusively of atoms of a single kind.

Identifying such 'elemental' substances was unlikely to be easy. After all, the whole basis of Democritus' atomic thesis was that the complexity of the world reflected the endless combinations of its basic building blocks. The likelihood was therefore that most elemental substances were bound together with other elemental substances and that very few were actually in their pure state.

The Greeks had considered the primary constituents of the world to be water, air, earth and fire. However, in reality, none of these, apart from water, was even close to being elemental. It would be left to others, equally wrong in their beliefs, to inadvertently identify the real primary constituents of matter. These were the alchemists who, during the Middle Ages, struggled heroically to 'transmute' base substances like lead into precious substances like gold. In the process, the alchemists accumulated a wealth of information about how substances combined with each other. In attaining their stated goal, however, they failed utterly. It was impossible to turn lead into gold. But this in itself was an important discovery, had the alchemists only recognised it. It was a strong indication that some substances were truly permanent and indestructible. All that was needed was for someone to draw the right conclusion. The man who did so was Antoine Lavoisier, a French aristocrat whose life was ended by the guillotine in the spring of 1794.

Five years before his death, Lavoisier compiled the first list of substances which he believed could not, by any means, be broken down into simpler substances. Lavoisier's list consisted of 23 'elements'. Some later turned out not to be elements at all but many were indeed elemental. They included sulphur and mercury, iron and zinc, silver and gold. Lavoisier's scheme was a turning point in the history of science. It signalled the death of alchemy and the birth of chemistry. The practitioners of the new science took as their starting point the existence of nature's elements and sought to combine these into new patterns. In doing so, they created 'compound' substances which had never before existed in the world. For chemists everything was in the combinations. And, because of the endless number of ways of combining nature's elements, in twos and threes and fours and so on, chemistry was a science with infinite possibilities.

In all likelihood each of Lavoisier's elements was a great mass of one kind of atom. However, the French chemist did not explicitly connect the concept of elements with the concept of atoms. This was left to an English schoolmaster and amateur scientist called John Dalton. In 1803, Dalton noticed that when elements combined to make a compound, they always did so in fixed proportions. For instance, when oxygen and hydrogen united to make water, precisely 8 grams of oxygen was used up for every 1 gram of hydrogen.[*] It was Dalton's genius to see in this simple observation the unmistakable fingerprint of invisible atoms combining with each other.

The observation was exactly what you would expect, Dalton reasoned, if oxygen consisted of large numbers of oxygen atoms, all identical, and hydrogen large numbers of hydrogen atoms, again all identical, and that the formation of water from oxygen and hydrogen

---

[*] These are the modern values. Dalton, for all his theoretical brilliance, was a sloppy experimenter and his measurements were poor even by the standards of his time.

involved the two kinds of atoms colliding and sticking to make large numbers of particles of water. Today, we call such particles 'molecules'. Since water has an identity as distinctive as either oxygen or hydrogen, it followed that water molecules were all identical. In other words, they each contained a fixed number of oxygen atoms and a fixed number of hydrogen atoms. Now if oxygen atoms all had a certain weight which was unique to oxygen and hydrogen atoms had a certain weight which was unique to hydrogen, then a fixed number of oxygen atoms translated into a fixed weight of oxygen atoms and a fixed number of hydrogen atoms translated into a fixed weight of hydrogen atoms. Each water molecule must therefore contain the same weight of oxygen atoms relative to hydrogen atoms.

Here then was the reason why the 'law of fixed proportions' applied to water. It was merely a reflection of the fact that each molecule of water contained a fixed number of oxygen atoms and a fixed number of hydrogen atoms.

If, say, the oxygen atoms in a single water molecule weighed 8 times as much as its hydrogen atoms, then the oxygen atoms in a million water molecules would still weigh 8 times as much as the hydrogen atoms in a million water molecules. It was irrelevant how much water was involved – the same factor would always hold. The observation that water used up 8 grams of oxygen for every gram of hydrogen therefore indicated that the oxygen atoms in a single water molecule weighed 8 times as much as the hydrogen atoms.

Dalton hazarded a guess that each water molecule contained just one oxygen atom bound to one hydrogen atom. It enabled him to conclude that an oxygen atom must weigh 8 times as much as a hydrogen atom. He was wrong. Today, everyone knows that the formula for water is $H_2O$ and that each water molecule in fact contains two atoms of hydrogen and one atom of oxygen. Rather than being 8 times as heavy as a hydrogen atom, an oxygen atom is actually 16 times as heavy. However, this minor error affected none of Dalton's reasoning.

The law of fixed proportions holds because a compound consists of a large number of identical molecules, each made of a fixed number of atoms of each component element. Just as Bernoulli had seen the unmistakable fingerprint of the atom in motion in the behaviour of a gas, Dalton had seen the fingerprint of the interacting atom in the way elements combined with each other.

Now two entirely different lines of reasoning had yielded independent evidence of atoms. Everything in the garden seemed rosy. However, there was the small matter of the number of different elements and, by implication, the number of different kinds of atom.

Democritus had never specified how many distinct types of atom there should be. However, his entire thesis had been that the complexity of the world was a consequence of the combinations of a limited number of fundamental building blocks. Lavoisier's list of elements indicated that there were about 20 different kinds of atom. However, the number of elements proliferated and another 32 were added to the list in the forty years after the French chemist's death.

Fifty kinds of fundamental building block seemed rather excessive. Why did nature have so many? In 1815, one man came up with a sensational answer. His name was William Prout, and he was the first to suggest that atoms were not the smallest things.

# 2

# *Atoms Are Not the Smallest Things*

HOW WE DISCOVERED THAT ATOMS ARE MADE
OF SMALLER THINGS AND THAT HEAVY ATOMS
CAN CHANGE INTO LIGHTER ATOMS,
UNLEASHING A DAMBURST OF ENERGY.

> Now make a fist, and if your fist is as big as the nucleus of an atom then
> the atom is as big as St Paul's, and if it happens to be a hydrogen atom
> then it has a single electron flitting about like a moth in an empty
> cathedral, now by the dome, now by the altar.
>
> Tom Stoppard

William Prout was an English physician with a fascination for the fledgling science of chemistry. In a laboratory at his London home he neatly combined his two interests by pioneering the study of gastric juice, urine and other body chemicals. Prout's enduring fame, however, rests on a startling claim he made about atoms in 1815.

Prout had compared the weights of different atoms using a controversial method originated by the Italian scientist Amedio Avogadro. In 1811, Avogadro had claimed that the same volume of any gas under identical conditions of temperature and pressure would always contain the same number of fundamental particles, or molecules. For instance, a litre of hydrogen would contain exactly the same number of molecules as a litre of oxygen or a litre of carbon dioxide.

It was a bizarre claim, and most chemists would not come round to accepting it for at least half a century. However, for those like Prout who were willing to suspend their disbelief, Avogadro's hypothesis provided a way to compare the weights of different molecules. It was only necessary to weigh equal volumes of different gases and compare them. Since equal volumes contained equal numbers of molecules, the comparison was exactly the same as comparing the weights of the individual molecules of each gas.

In fact, it was possible to do even better than this, thanks to a discovery by the French scientist André Ampère. In 1814 Ampère had found that if a gas consisted of a single element, its atoms invariably clumped in pairs. For instance, the molecules of oxygen consisted of pairs of oxygen atoms; the molecules of chlorine, pairs of chlorine atoms. Comparing the weights of equal volumes of such gases was therefore the same as comparing the weights of the atoms of each gas.

Using this method of atomic comparison, Prout very soon stumbled on something remarkable: all atoms appeared to have weights which were exact multiples of the weight of hydrogen, the lightest atom. For instance, nitrogen was exactly 14 times as heavy as hydrogen and oxygen exactly 16 times as heavy as hydrogen. Finding such regularity was astonishing enough. However, it was nowhere near as astonishing as Prout's interpretation of the regularity. His observations, he maintained, were unmistakable proof that all atoms were actually made out of hydrogen atoms.

It was an extraordinary claim. It struck at the very basis of chemistry, founded as it was on the indestructibility of the elements. For if it were accepted that all atoms were built from hydrogen atoms, then clearly it must be possible to transform an atom of one element into an atom of another. The spectre of alchemy, so successfully exorcised by Lavoisier and Dalton, would be back to haunt science.

Prout's claim not only revived forlorn hopes that lead might be transmuted into gold. It also struck at the very heart of the atomic idea. For if atoms had been assembled from other things, it stood to reason that they themselves were not the smallest things in creation. If true, Prout's hypothesis dealt a death blow to the 2000-year-old idea that atoms were the ultimate building blocks of the world.

For most scientists, however, reports of the death of atoms were premature. Not only was Prout's method of weighing atoms highly controversial; other chemists, stimulated by Prout's work, soon discovered exceptions to Prout's rule. The weights of some elements were very definitely not exact multiples of the weight of hydrogen. A prime example was chlorine, which was 35.5 times as heavy as hydrogen. A rule that applies only part of the time is difficult to take seriously. The fate of Prout's was to languish in a backwater of science for more than a century. In the meantime, there were other, tantalising hints that atoms were indeed made of smaller things.

The most striking came from Dimitri Mendeleev, a chemist from

Siberia whose other claims to fame were that he was a bigamist and that he had once crossed swords with Nikolay Tolstoy. Mendeleev spotted a curious repeating pattern in the properties of the elements. The pattern had become apparent during the writing of a textbook. Mendeleev had prepared a series of cards, each listing the main properties of an element. He was suddenly struck by the fact that if the cards for the 67 known elements were placed in horizontal rows, with most of the elements in order of increasing atomic weight, then elements with similar properties appeared in vertical columns.* This arrangement of the elements, with its characteristic rows and columns, became known as the Periodic Table. Using it, Mendeleev was able to predict the properties of 'missing' elements, which were then in turn discovered.

A repeating pattern in natural phenomena is a strong indication that there exists a simpler, more compact way of describing them. In the case of the Periodic Table, the pattern suggested that the 67 distinct atoms could be described in terms of significantly fewer than 67 'subatomic' building blocks. Atoms, in short, were made of smaller things.

But although the hint was there, neither Mendeleev or any of his contemporaries could make the slightest sense of the pattern in the Periodic Table. It was destined to remain an enigma for another half century. The definitive evidence that atoms were made of smaller things came only when the atom began suddenly and dramatically to fall apart. It began with a discovery made by a French chemist, Henri Becquerel, in 1896.

## THE DISCOVERY OF URANIUM RAYS

Becquerel was a man with an obsession. Like his father and grandfather before him, both of whom had held his chair in physics at the Ecole Polytechnique in Paris, he was intrigued by the peculiar ability of some minerals to glow, or 'fluoresce', after exposure to sunlight. Becquerel was investigating the phenomenon of fluorescence in January 1896 when the dramatic news reached Paris of the discovery of X-rays.

An obscure German physicist called Wilhelm Roentgen had observed the rays emerging from a cathode ray tube — a glass tube evacuated of air and closely related to a modern TV tube. The rays stabbed through solid matter as if it were thin air. Roentgen had used them to 'photograph' the bones inside his wife's hand and the picture, reproduced in a thousand newspapers across the world, had created a sensation. But it

---

* Today, more than 100 elements are known, although only 92 are naturally occurring.

wasn't the picture that caused Becquerel to sit up. It was the report that Roentgen's X-rays had come from a fluorescent spot on the wall of his cathode ray tube. Becquerel was struck by the sudden thought that the fluorescent minerals he had long been studying might be glowing with invisible X-rays as well as ordinary light.

He was barking up the wrong tree. X-rays had nothing whatsoever to do with fluorescence. Nevertheless, Becquerel was led by his faulty logic to embark on a programme to test large numbers of fluorescent minerals to see whether any were giving out X-rays. The technique he used was simple. He wrapped a photographic plate in dark paper so that light could not get at it and placed it on a sunlit windowsill for several hours. On top of the plate he arranged various minerals which were known to fluoresce. If sunlight triggered a mineral to produce X-rays in addition to visible light, reasoned Becquerel, the X-rays should easily penetrate the paper and blacken the photographic plate.

Sadly, when he developed his plates, he observed no such effect. However, after weeks of disappointment, Becquerel came to a sample of the uranium 'salt', potassium uranyl disulphate. A colleague had borrowed it from him and only just returned it. Becquerel had high hopes for the salt. His grandfather had found that, of all fluorescent minerals, those which contained uranium – an element noteworthy only because it was the heaviest in nature – invariably produced the most brilliant effect.

Becquerel's hopes were amply fulfilled. On 24 February 1896 he reported to the French Academy of Sciences that his uranium mineral did indeed emit rays that penetrated paper and blackened a photographic plate. Such was his cautious nature, however, that he never once mentioned his suspicion that the penetrating rays might be X-rays.

Becquerel next set about determining the properties of the rays coming from the uranium salt. He quickly devised another experiment in which a small cross of copper was interposed between the sample and the wrapped photographic plate. If the rays travelled in straight lines, like Roentgen's X-rays, then when the plate was developed it would show the shadowed outline of the copper cross.

On 26 February, much to Becquerel's frustration, the sky above Paris was completely overcast and he was unable to carry out his experiment. Instead, he took his entire apparatus – uranium salt, wrapped photographic plate and copper cross – and placed it in the drawer of a cabinet. There it remained, in total darkness, for several days during which time the sun made only the most fleeting of appearances in the wintry sky above the city. On Sunday 1 March, however, with the cloud cover over Paris as thick as ever, Becquerel's impatience got the better of him. He removed his apparatus from the dark drawer and developed the photographic plate.

Why he decided to do this when he was looking specifically for an effect triggered by sunlight, and he knew full well that the plate had languished for days in utter darkness, is a fascinating question. Perhaps he had a hunch. Perhaps it was a sixth sense, the flash of unpredictable genius that separates the few scientists who make great discoveries from the many who do not. Whatever the truth, Becquerel developed the plate. And what he saw when he did so left him open-mouthed with incredulity.

Shining out, brilliant white, against the black background was the image of the copper cross.

On the street outside Becquerel's laboratory, churchgoers hurried along, summoned by the sound of ringing bells. Horse-drawn carriages clattered over the cobbles. A dog barked, a few forlorn birds flew across the watery sky. The world was the same as it had been an instant earlier. Yet the world had changed irrevocably. The road that led to Hiroshima and Chernobyl would one day be traced back to this moment, this place.

Becquerel could not tear his eyes from the shadow of the cross. The rays he had reported to the Academy of Sciences barely a week before were still there, undiminished in intensity. Yet, in the meantime, not the slightest drop of sunlight had reached the uranium. There was only one explanation. The rays coming from the potassium uranyl disulphate were not triggered by sunlight or by any other obvious external agent. Instead, they were intrinsic to the uranium salt itself. What Becquerel had discovered was an entirely new phenomenon – one which, two years later, would be christened 'radioactivity'.

The most remarkable characteristic of radioactivity was its persistence. As time passed, Becquerel could detect not the slightest weakening in the 'uranium rays'. Week after week, month after month, the rays flooded forth in an unending stream, drawing on an apparently bottomless well of energy.

## THE DISINTEGRATING ATOM

Becquerel's discovery caused nothing of the stir of Roentgen's and only a handful of scientists took up the challenge to understand the phenomenon. One of them was the Polish-French chemist Marie Curie, who at the time was looking for a research topic for a doctoral thesis. It was she, in fact, who would coin the term radioactivity. Curie discovered that Becquerel's rays were emitted not only by uranium minerals but also by minerals containing thorium, the second heaviest element known. In the course of investigating both types of minerals she then made an even greater discovery. Two minerals – pitchblende and chalcite – were far more radioactive than their content of uranium and thorium implied. Her daring explanation was that the minerals contained radioactive elements as yet unknown to science.

Marie and her husband Pierre began a painstaking search for the new elements. In 1898, by the truly herculean effort of dissolving, filtering and crystallising many tonnes of the pitchblende they separated out minuscule quantities of two new elements, which they christened polonium and radium. Radium was more than a million times as radioactive as uranium and it glowed threateningly in the dark. It would make radioactivity known to the whole world and Marie Curie the most famous women scientist in history.

Marie Curie's success in discovering two new elements by their tell-tale radioactivity was powerful proof that radioactivity was a property of atoms. But what was it telling us about atoms? Curie herself would not find the answer. That was left to the son of a poor New Zealand farmer called Ernest Rutherford.

Rutherford, although born in one of the most remote regions of New Zealand, managed to get a good education and a place at the university in Christchurch. There, he demonstrated his genius for experiment by building a device for detecting radio waves at a distance which anticipated Marconi. It was impressive enough to earn him a scholarship to Cambridge.* The 25-year-old Rutherford had been at Cambridge less than a month when X-rays were discovered. He immediately embarked on an investigation of the phenomenon. It was only much later, when the X-ray work was completed, that Rutherford turned his attention to the new puzzle: radioactivity. In fact, his experiments were carried out not at Cambridge but at McGill University in Montreal, where Rutherford accepted a professorship in 1898. There, he discovered that radioactive atoms gave out not one but two distinct types of rays, which he named alpha and beta rays.† What

---

* The scholarship had in fact been awarded to another man, ranked above Rutherford, but the man had chosen to get married. In later years, when he stood at the very pinnacle of British science, Lord Rutherford, the greatest experimental scientist of his generation, would break down and cry at the thought of how the course of his life could so easily have gone another way.

† In fact, a third type was discovered in 1900 by the French physicist Paul Villard and named gamma rays. However, gamma rays – which would eventually be recognised as a kind of X-rays – were destined to play only a minor role in illuminating radioactivity.

distinguished the two kinds of ray was their ability to stab through solid matter. Alpha rays were far less penetrating than beta rays.

An important step towards solving the puzzle of alpha rays was taken in 1903 by two chemists, the English Frederick Soddy and William Ramsay, who was Scottish. In 1895, Ramsay had discovered helium, the lighter-than-air gas that fills Mickey Mouse balloons and makes your voice sound like Donald Duck. He had found it in the gases given off from a uranium mineral called cleveite. What Ramsay could not possibly have known — because Becquerel's discovery of uranium rays was still a year in the future — was that cleveite was radioactive. Once Becquerel had announced his discovery, the fact that helium had been found in a uranium-bearing mineral raised a few eyebrows. Was it just a coincidence that helium was associated with radioactivity? Or was there a connection?

In 1903, Ramsay and Soddy, working at University College in London, collected a tiny quantity of the gas produced by a radium salt and demonstrated that it was helium. It became clear that the association between helium and radioactivity was no accident. Helium was created by radioactivity. In 1908, Rutherford made the link even more explicit. By now back from Canada and working at the University of Manchester with a young German physicist called Hans Geiger, he managed to collect alpha rays emitted by radium and show that they were indeed the gas helium.[*]

All doubt was gone. Alpha rays were helium atoms.[†] During the process of radioactivity, one kind of atom — radium — was actually spitting out another kind of atom — helium. Helium was a sub-atom. It was a dramatic demonstration that atoms were indeed made of smaller things.

The discovery that alpha rays were helium atoms also shed light on the nature of radioactivity. For an atom which spat out another kind of atom must necessarily be changed in the process. It was the confirmation of a revolutionary idea which had been conceived by Rutherford and Soddy between 1901 and 1903. At McGill the two men had found increasing evidence that a radioactive atom is simply a heavy atom which happens to be unstable. Inevitably — after a second or a year or a million years — it disintegrates by expelling an alpha, beta or gamma ray. What remains after such a 'decay' is an atom of a slightly lighter element.

A radioactive atom is therefore an atom changing itself into another

---

[*] Geiger is famous for the Geiger counter, whose machine-gun spluttering warns of high levels of radioactivity.

[†] In fact, they were positively charged helium atoms. They became true helium atoms only when they slowed down and their charge was neutralised by picking up electrons.

kind of atom. And, in its desperate quest for stability, it may decay more than once. For instance, uranium transforms itself into a succession of lighter and lighter atoms (one of which, incidentally, is radium). Only after the emission of its eighth alpha particle does it finally find peace and stability as a non-radioactive atom of lead. Rutherford and Soddy's theory of 'atomic transmutation' was deeply disturbing to the scientific community. It was a revolutionary break with the past and the idea of indestructible atoms. Nothing appeared sacred any more. Everything in the everyday world, even the building blocks of our bodies, might succumb to the vagaries of atomic disintegration.

But radioactivity was not alone in pointing to an atom made of smaller things. There was other evidence. It was provided by a British physicist called Joseph John Thomson who was investigating the phenomenon of 'cathode rays'.

## A CHIP OFF THE OLD ATOM

Cathode rays had been discovered in 1858 by Julius Plücker, a physicist at the University of Bonn. Like many scientists of his day, Plücker was intrigued by electricity. The problem was that electricity was inextricably mixed up with the material that was carrying it. When flowing through a copper wire, for instance, its properties were hopelessly bound up with the properties of the copper. What was needed was a way to observe electricity on its own.

Some progress had in fact been made more than a century earlier by Francis Hauksbee, a demonstrator of scientific experiments at the Royal Society in London. In 1709, Hauksbee had discharged electricity through a glass jar after first pumping out most of the air. But even though the air in the jar was 60 times thinner than normal it still interfered with the free flow of electricity, forcing it to zigzag like lightning around the air molecules in its path. If electricity were to be observed in its naked state, even thinner air would be needed. This, in turn, would require a better air pump than the one used by Hauksbee.

Such a device was invented in 1855 by Johann Geissler, a mechanic and glassblower from Bonn. With his 'mercury air pump', Geissler was able to suck the air out of a glass vessel until what remained was almost 10,000 times thinner than normal – so rarefied it was virtually indistinguishable from empty space. It was perfect for studying electricity in the raw, a fact recognised immediately by Plücker, whose daily walk to and from the University of Bonn took him past Geissler's workshop. Plücker commissioned Geissler to make him a long, thin glass tube with two metal plates, or 'electrodes', sealed inside. The electrodes, one at either end, protruded through the glass wall so they could be connected to a source of outside electricity.

When Hauksbee had discharged electricity through his glass jar, the thin air had glowed with an eerie light. When electricity leapt from electrode to electrode inside Plücker's 'discharge tube' the glow was even more spectacular. In fact, it was possible to make it any colour of the rainbow by using different rarefied gases in the tube, a fact soon recognised by opticians, who put the tubes in shop windows to amaze their customers. They glowed like modern-day fluorescent lights or television sets, pieces of twentieth-century technology which by some fluke of time had inadvertently fallen into the middle of the nineteenth century.

In 1858, Plücker began using his discharge tube to investigate how well rarefied gases conducted electricity. Very quickly he turned up an odd result. When he pumped out almost all the air from his tube, the eerie glow which had filled its interior disappeared – but the tube did not go completely dark. On the glass near one of the electrodes there appeared a peculiar greenish glow. Plücker's student Johann Hittorf showed that the glow was caused by invisible 'rays' which stabbed outwards from the other electrode. The rays cast shadows when objects were placed in their way, indicating they travelled in straight lines; and they were deflected by a magnet, indicating they were electrically charged. The greenish light evidently marked the spot where the rays slammed into the wall of the tube. Hittorf christened the rays 'glow rays', but the name which stuck was 'cathode rays' since the rays came from the electrode known as the 'cathode'.*

But what were cathode rays? As the nineteenth century drew to a close, there were two opposing views. One, promoted by German scientists, was that cathode rays were some kind of 'wave in space' not unlike a light wave. The other view, championed by British scientists, was that cathode rays were streams of tiny negatively charged particles.

After thirty years of debate, the nature of cathode rays was finally elucidated by the British physicist Joseph John Thomson, or 'J. J.' as he was widely known. He was the man who had employed Rutherford as his student when the New Zealander had first arrived at Cambridge. Rutherford's investigation of X-rays had been carried out under the

---

* The cathode was the source of 'negative' electricity, in contrast with the anode which was the source of 'positive' electricity. The fact that electricity came in two kinds – negative and positive – had been known ever since the eighteenth century. The American physicist Benjamin Franklin was the first to recognise that electricity was a single 'fluid', with an excess manifesting itself as positive electricity and a deficit as negative electricity.

supervision of the older man. Thomson was determined to get to the bottom of the cathode ray mystery. In 1897, he drilled a hole in the anode of a discharge tube to allow the mysterious rays from the cathode to stab right on through. In the space after the anode, he arranged that the magnetic force field from a magnet would tug the cathode rays in one direction and the electric force field between two electrically charged metal plates would tug them in the opposite direction. After being fought over in this way, the rays would finally strike the glass wall of the tube to create a familiar greenish spot of light.

The key to Thomson's experiment was the glowing spot, whose position indicated how much the beam of cathode rays had been deflected. The deflection could be made zero by adjusting the magnetic and electric forces so that they perfectly balanced. In such a situation, Thomson could read off the strength of the electric force. He also knew in theory how the magnetic force on charged particles depended on their speed. By equating the two forces, he was therefore able to deduce the one thing he did not know — the speed of the cathode rays.

Knowing the speed of the cathode rays was important because it was one factor which influenced how wildly they were deflected when subjected to an electric force. The slower the cathode ray particles, the longer they were exposed to the electric force and the greater the deflection of the glowing dot. The deflection was also influenced by two unknown factors: the electric charge carried by the cathode ray particles, and their mass. The larger the charge, the greater the electric force the particles felt and the greater their deflection. The smaller the mass, the easier it was for any force to push the particles about and, again, the greater their deflection.

Knowing the deflection of the glowing dot, and the velocity of the cathode ray particles, Thomson expected to be able to deduce their charge and mass. In fact, he did not do quite as well as this. What he actually deduced was a combination of their charge and mass. Fortunately, however, there was independent evidence from electrolysis — the passing of electricity through liquids — that electric charge came in discrete chunks. By assuming that such chunks were carried by individual cathode ray particles, Thomson was able to deduce their mass. The figure he came up with was a thousand billion billion billionth of a kilogram — or a mere 2000th of the mass of a hydrogen atom.

Atoms were made of smaller things. But the fundamental building blocks of matter were not hydrogen atoms, as Prout had maintained, but vastly smaller entities. Thomson's particles were christened 'electrons'. They were the elusive particles of electricity. Ripped free from atoms, they sailed through space as cathode rays, or drifted along a copper wire as an electric current.* The electron was the first 'subatomic' particle. And Thomson used it to create the first scientific picture of the atom. What he visualised was a multitude of tiny electrons embedded 'like raisins in a plum pudding' in a diffuse ball of positive charge.

It is a characteristic of electrical charge that unlike charges attract each other whereas like charges repel. The atom would therefore be held together by the attractive force between the negatively charged electrons and the diffuse ball of positive charge. Thomson's 'plum pudding model' was the accepted picture of the atom at the start of the twentieth century. But, in 1911, it was blown out of the water when, incredibly, Rutherford found a way to probe deep inside the atom.

## INSIDE THE ATOM

A typical atom was a mere 10 billionth of a metre across. The building blocks of matter were so mind-cringingly small that it seemed impossible that anyone would ever be able to see inside one. It was a tribute to the experimental genius of Ernest Rutherford that he discovered a way.

Rutherford's idea was to fire tiny projectiles into them and observe how they were deflected by the 'subatomic' structures they encountered. Working backwards from the pattern of the deflections, it might be possible to deduce something of the internal structure of the atom. For the scheme to work, the projectiles had to be smaller than atoms. They also had to slam into atoms with the violence of subatomic express trains in order to penetrate deep into their cores. Rutherford's brainwave was to realise that nature had provided the perfect projectiles: alpha particles.

Since alpha particles actually came from inside atoms, Rutherford's proposal boiled down to using the atom to reveal its own structure. It was the kind of bare-faced audacity that so typified Rutherford's assault on nature. He exploited every weapon at his disposal, often in the most disarmingly simple experiments. And those experiments did not come much simpler than the one in which he fired alpha particles at atoms to see how they would rebound.

Rutherford first attempted the trick in Montreal in 1906. But he delegated the task to two younger physicists – Ernest Marsden and Hans Geiger – when he moved to Manchester the following year. Marsden was a fellow New Zealander. Geiger was the German physicist who in 1908 would help Rutherford prove that alpha particles and helium

---

* In fact, as Henri Becquerel discovered, electrons were also spat out by atoms as beta rays.

atoms were one and the same. Geiger and Marsden's 'alpha–scattering' experiment used a small sample of radium, which fired off alpha particles like microscopic firecrackers. The sample was placed behind a lead screen containing a narrow slit, so that a thread-thin stream of alpha particles emerged on the far side. It was the world's smallest machine-gun, rattling out subatomic bullets.

In the firing line Geiger and Marsden placed a gold foil only a few thousand atoms thick. It was insubstantial enough that all the alpha particles from the miniature machine-gun would pass through. On the other hand, it was sufficiently substantial to ensure that some would pass close enough to gold atoms to be deflected slightly from their path. The effect of many such deflections should be to cause the narrow beam of alpha particles to fan out, and it was this fanning out that Geiger and Marsden intended to measure with a special screen placed in the path of the emerging beam. The screen was painted with zinc sulphide, a chemical known to 'scintillate', or emit a tiny flash of light, each time it was struck by an alpha particle.

The flashes teetered on the edge of invisibility. However, by squinting through a microscope in a pitch-black laboratory, Geiger and Marsden could just about make them out. The cluster of tiny stars that twinkled on the zinc sulphide screen revealed that the beam of alpha particles had indeed fanned out after passing through the foil. And the effect, as expected, was slight. No individual particle was deflected from its path by more than a few degrees.

The 'alpha–scattering' experiment might have ended at this point had it not been for an outrageous suggestion made by Rutherford when passing Geiger and Marsden's laboratory one day in 1909. The suggestion was that they look to see whether the foil was deflecting any alpha particles by very large angles. That such a thing might happen was inconceivable. According to Thomson's plum pudding model, an atom consisted of a multitude of pin-prick electrons embedded in a diffuse globe of positive charge. The alpha particles Geiger and Marsden were firing into this flimsy arrangement, however, were unstoppable subatomic express trains, each as heavy as around 8000 electrons. The chance of such a massive particle being seriously deflected by an electron was about as great as that of a real express train being derailed by an old lady on a bicycle.

Rutherford knew this as well as anyone. However, he had the nerve to suggest that Geiger and Marsden look for the impossible. Later, he would claim that he had made his suggestion simply 'for the sheer hell of it', and that he had not really expected to see a single alpha particle deflected through a large angle. However, Rutherford must have had a hunch – even a faint hunch – that something unusual might turn up in the alpha-scattering experiment. If all his successes in science had taught him one lesson, it was the crucial importance of keeping his mind open for the unexpected, and of being perpetually prepared to be surprised by nature. And never did nature surprise him more than when, three days later, Geiger burst into his office with the unbelievable news that one in every 8000 alpha particles was actually bouncing backwards from the gold foil. 'It was quite the most incredible event that has ever happened to me in my life,' Rutherford later admitted. 'It was almost as incredible as if you fired a 15-inch shell at a piece of tissue paper and it came back and hit you.'

There was simply no way the plum pudding model could explain this observation. An atom was not a flimsy thing at all. Something buried deep inside could stop an alpha particle dead in its tracks and turn it around. That something could only be a tiny nugget of positive charge, sitting at the dead centre of an atom and repelling the incoming positive charge. Since the nugget was capable of standing up to a massive alpha particle without being knocked to kingdom come, it too must be massive. In fact, it must contain almost all of the mass of an atom. Rutherford had discovered the atomic 'nucleus'.

## THE CATHEDRAL OF THE ATOM

In fact, the idea of the nucleus was born after a gestation of more than a year, in which Rutherford discarded many alternative explanations for Geiger and Marsden's baffling result. In the end, however, an atom with a massive, positively charged nucleus made the most sense. Rutherford's atom was a miniature solar system, in which negatively charged electrons were attracted to the positive charge of the nucleus and orbited it like planets round the sun. The mass of the nucleus had to be least as great as an alpha particle – and probably a lot more – in order that the nucleus not be kicked out of the atom. It therefore had to contain more than 99.9 per cent of the mass of an atom. The nucleus also had to be very, very tiny. Only if nature crammed a large positive charge into a very small volume could a nucleus exert a repulsive force so overwhelming that it could make an alpha particle execute a U-turn.

The nucleus was Rutherford's greatest triumph. Alone of the pioneers of radioactivity, he continued to make major discoveries in the field. As the physicist and novelist C. P. Snow observed: 'As soon as Rutherford got on to radioactivity, he was set on his life's work. His ideas were simple, rugged, material; he kept them so. He thought of atoms as though they were tennis balls. He discovered particles smaller than atoms, and discovered how they moved or bounced. Sometimes the particles bounced the wrong way. Then he inspected the facts and made a new but always simple picture. In that way, he moved, as certainly as a sleepwalker, from unstable radioactive atoms to the discovery of the nucleus and the structure of the atom.'

What was most striking about Rutherford's vision of an atom was its appalling emptiness. Electrons were like Stoppard's moths in the empty dome of St Paul's Cathedral, flitting about a central nucleus which was hardly bigger than a clenched fist. The staggering implication of this was that the familiar world, despite its appearance of solidity, was in fact no more substantial than a ghost. Matter, whether in a chair, a human being or a star, was almost exclusively empty space. What substance an atom possessed resided in its impossibly small nucleus.

But what was an atomic nucleus made of? This now was the burning question. Unfortunately, the nucleus was about 100,000 times smaller than an atom, and it would take two decades of painstaking experiment to come up with an answer.

## INSIDE THE ATOMIC NUCLEUS

The picture that gradually emerged was of a nucleus that contained a very heavy particle with a positive electric charge. The particle, christened the 'proton', was none other than the hydrogen 'building block' envisaged by William Prout a century earlier. It was almost 2000 times heavier than an electron, and it carried a charge which was equal but opposite to the charge on the electron. In the lightest of all atoms, hydrogen, the positive charge of a single proton in the nucleus exactly balanced the negative charge of a solitary electron in distant orbit.

The key feature that distinguished a hydrogen atom from, say, an atom of carbon or uranium, was the number of protons in its nucleus. It was this that determined its properties. Whereas hydrogen had a single proton and a single electron, carbon had six protons and six electrons. The nucleus of uranium, on the other hand, was a monster with 92 protons and sat in the midst of a haze of 92 whirling electrons. The protons in a nucleus were always balanced by an equal number of electrons so that the atom was kept electrically neutral.

This two-particle picture of an atom was pleasingly simple. However, its deficiencies were glaringly apparent when it came to describing even the second lightest atom: helium. Rutherford and Geiger had shown that a helium atom and an alpha particle were the same, apart from the positive charge carried by an alpha particle. After the discovery of the atomic nucleus, it was clear why. An alpha particle was a helium nucleus. As it gradually ploughed to a halt after being spat out by a radioactive atom, it mopped up electrons along its path. These neutralised its positive charge and transformed it into a helium atom. Since the charge carried by an alpha particle was twice the charge on a proton, two electrons were required to do this. Physicists were led inexorably to the conclusion that a helium atom consisted of a pair of electrons circling a nucleus containing a pair of protons.

The trouble was this did not square with the weight of the helium nucleus. Since the nucleus contained two protons, it was reasonable to expect that it would weigh the same as two protons. Confounding all expectations, however, a helium nucleus weighed as much as four protons. And the weight discrepancy grew worse for atoms heavier than helium. In general, a nucleus was more than twice as heavy as the protons it contained. Clearly, there must be other things besides protons in an atomic nucleus. Initially, Rutherford suspected the other things were electrons. By mixing them judiciously with protons, it was possible to reproduce the mass and charge combination of any known nucleus. For instance, two electrons and four protons could make a nucleus of helium. The negative charge of the electrons would cancel out the positive charge of two of the protons, creating a nucleus with the electric charge of two protons but the mass of four.

Having electrons inside an atomic nucleus as well as in orbit round it initially made a great deal of sense. After all, beta particles were electrons and, by 1913, Rutherford had become convinced they originated in the nucleus just like alpha particles. What was more reasonable than that beta decay was the expulsion of an electron already in a nucleus? However, it turned out that the electron was far too slippery a customer to be confined in the prison of a nucleus. By 1920, Rutherford had abandoned the idea of nuclear electrons and was proposing a startling new alternative.

The alternative was that the nucleus contained an entirely new particle – one as heavy as a proton but with no electrical charge. Rutherford dubbed the massive neutral particle a 'neutron'. It was a brilliant guess. However, Rutherford was way ahead of his time and it was not until 1932 that the existence of the neutron was finally confirmed by Rutherford's Cambridge protégé, James Chadwick. Rutherford's only mistake was to think of the neutron as a composite particle – a proton and an electron sandwiched together. The neutron turned out to be a particle in its own right, every bit as fundamental as either the electron or the proton.

The discovery of the neutron made immediate sense of an observation that had puzzled people from the earliest days of radioactivity. This was that many elements came in a variety of radioactive forms, each with different radioactive properties such as a different decay rate. A prime example was lead. There was a common form that was unradioactive and a rare form, found in uranium–bearing minerals, that was radioactive. The difference between the various radioactive forms of an element soon became clear: each consisted of atoms with a slightly different mass. The variants were christened 'isotopes' by Frederick Soddy in 1911.

In fact, the idea that each element might be a mixture of atoms of different weights had been proposed as early as 1871 by Sir William Crookes, one of the pioneers of the cathode ray tube. It explained why Prout's rule was violated by a handful of elements whose atomic weights were not simple multiples of the weight of hydrogen. The most notorious example was chlorine, which weighed 35.5 times that of hydrogen. According to Crookes, such anomalous elements consisted of a mixture of atoms of different weights. What chemists were therefore measuring when they weighed such an element was the average weight of all its atoms. This average was not restricted to a multiple of the weight of hydrogen so could take on any value whatsoever. It took radioactivity to breathe new life into Crookes' old idea because most of the variant forms of each element turned out to be radioactively unstable. An element like chlorine, with more than one stable isotope, was rare.

The discovery of the neutron at last shed light on the nature of isotopes. The various isotopes of an element were merely atoms with the same number of protons in their nucleus but a different number of neutrons. All atoms of chlorine, for instance, had a complement of 17 protons. However, nature must permit the element to have more than one isotope, each with a different number of neutrons.

The neutron triumphantly completed the simple picture of the atom. Every atom was assembled from just three building blocks: electrons, protons and neutrons. The protons and neutrons clung together in a tight central clump – the atomic nucleus – while the electrons circled in a distant haze. It was the neutrons which were responsible for increasing the weight of the elements without adding any electrical charge. Two protons and two neutrons made a helium nucleus; eight protons and eight neutrons an oxygen; 26 protons and 30 neutrons an iron nucleus; 79 protons and 118 neutrons a gold; and 92 protons and 146 neutrons a nucleus of uranium. When a radioactive nucleus expelled an alpha particle, it lost two neutrons and two protons and consequently became a nucleus of an element two places lower in the Periodic Table. When a radioactive nucleus emitted a beta particle, however, a neutron changed into a proton, transforming the nucleus into that of an element one place higher in the Periodic Table.

Whatever was powering these shuddering convulsions and sending alpha and beta particles rocketing outwards, it was now abundantly clear that it resided deep inside the atomic nucleus. The nucleus was where the action was. The nucleus was the seat of the energy which was set free during radioactive decay. Just how big that energy was was the real shock of radioactivity.

## THE EXTRAORDINARY ENERGY INSIDE ATOMS

There had been definite indications that a lot of energy was involved. In 1903, Rutherford had measured the speed of alpha particles expelled from radium atoms. He had found it to be 25,000 kilometres per second – a hundred thousand times faster than a present-day passenger airliner. A spacecraft travelling at such a speed could get to the moon in 16 seconds and the sun in an hour and a half. When radioactive atoms disintegrated, it was with an explosive violence that defied imagination.

But although an alpha particle packed a powerful punch on the atomic scale, its energy was so tiny on the human scale that measuring it directly was out of the question. On the other hand, a sample of radioactive material generated alpha particles in truly astonomical numbers. According to Rutherford's estimate, a single gram of radium unleashed more than 10 billion alpha particles every second. Measuring the total energy of all these should not be impossible. Alpha particles were the least penetrating of radioactive emanations – easily stopped by a sheet of paper. It meant that most of the alpha particles emitted by a sample of radioactive material were stopped within the sample itself. Their energy of motion was transferred to the atoms they collided with and, ultimately, ended up as heat. The radioactive sample got hotter. To measure the total energy liberated by alpha particles, it was therefore necessary only to measure how quickly a sample was generating heat.

The difficulty was that intensely radioactive material like radium was available in only the tiniest quantities, and measuring the heat generated was no mean feat. However, the problems were overcome in 1903, by Pierre Curie and his colleague Albert Laborde. What the two men discovered astonished them. The heat generated by a sample of radium was enough to melt the sample within a few hours if it were not allowed to escape. In fact, it was sufficient to heat a weight of water equivalent to the weight of radium from freezing point to boiling point in just 45 minutes. In other words, a kilogram of radium could boil a kilogram of water in 45 minutes. It could boil another kilogram of water in the next 45 minutes. And another and another. In fact, it could boil a kilogram of water every 45 minutes for thousands of years. Pound for pound, radium generated a million times as much energy as dynamite.[*]

Curie and Laborde's result generated huge international interest in radium. In his wonderful book, *Inward Bound*, the physicist-historian Abraham Pais tells of the display of a single grain of 'priceless mysterious radium' at the International Electrical Congress in St Louis in 1904. A journalist on the *St Louis Post Dispatch*, previewing the exhibition, wrote: 'Its power will be inconceivable. By means of the metal all the arsenals in the world would be destroyed. It could make war impossible by exhausting all the accumulated explosives in the world . . . It is even possible that an instrument might be invented which at the touch of a key would blow up the whole Earth and bring about the end of the world.'

Curie and Laborde's discovery was a bombshell dropped in the heart of the world of physics. Locked inside ordinary matter was an energy supply to dwarf the imagination. Where did the tremendous energy inside the atom come from? While the physicists scratched their heads and pondered the question, another group of scientists began to sit up and take notice. They were the astronomers. The ultimate origin of the energy inside atoms did not concern them. What had piqued their interest was the mere fact that such a prodigious source of energy existed. For it seemed to them that it had the potential to address one of the outstanding questions in astronomy. The question was: what makes the sun shine?

---

[*] Marie Curie's laboratory notebooks provide a sobering warning of the power of radioactivity. More than 60 years after her death, they are still considered too dangerous to handle and are kept, for safety reasons, in boxes lined with lead. When she flattened the pages, she flattened them with fingers swollen with radiation burns and impregnated with radioactive deposits. Even today, if a photographic film is placed between the pages and developed, the ghostly fingerprints of the discoverer of radium and polonium swim into view.

# Part Two
# Atoms and Starlight

# 3

# *The Disintegrating Sun*

HOW WE CAME TO REALISE THAT THE SUN MUST
BE DRAWING ON THE ENERGY LOCKED
INSIDE ATOMS.

The great mystery is to conceive how such an enormous conflagration as
the sun can be kept up. Every discovery in chemical science here leaves
us completely at a loss, or rather, seems to remove farther the prospect of
a probable explanation.

John Herschel

The maintenance of solar energy no longer presents any fundamental
difficulty if the internal energy of the component elements is considered
to be available – that is, if processes of subatomic change are going on.

Ernest Rutherford

From time immemorial people have wondered what makes the sun
shine. One of the first to come up with a plausible answer was the
Greek philosopher Anaxagoras. In the middle of the fifth century BC, he
proposed that the sun was 'a red–hot ball of iron not much bigger than
Greece'. It is impossible not to admire the audacity of Anaxagoras in
making such a confident suggestion. Not until 1658, when the Irish
bishop James Ussher claimed the earth was created on the evening of 22
October 4004 BC, did anyone couch a piece of guesswork more
precisely.

Unfortunately, Anaxagoras was a little wide of the mark. The sun is
not made of iron – although, bizarrely, most astronomers believed so
until well into the twentieth century – and the sun is considerably larger
than Greece, ancient or modern. However, the Greek philosopher can

be forgiven his error since the only course open to him was to extrapolate from the world of his direct experience. And who can deny that the sun looks like a giant ball of molten iron? If Anaxagoras can be faulted it is because he overlooked the fact that a ball of molten iron – even one slightly bigger than Greece – would inevitably cool and fade with time. The sun, on the other hand, was a paragon of constancy. There was absolutely no reason to believe that it had beamed down on the land of the first pharaohs any less brightly than it had on Anaxagoras' Greece.

Nowadays, of course, it is obvious to us that a hot body will cool if the heat leaking from it is not constantly replenished. Usually, this involves burning some kind of fuel. For instance, wood or coal is needed to keep a fire roaring on a winter's evening; a supply of carbohydrate-rich food to maintain our bodies at an optimum 37°C. However, this apparently simple intuition about the behaviour of hot bodies became common knowledge only a few centuries ago.

The first person to ask how the sun could have shone undiminished throughout the ages was Isaac Newton in the early eighteenth century. It was a question which had implications beyond the confines of our solar system because by Newton's time the stars were believed to be other suns. This view had originated with the sixteenth-century Italian philosopher Giordano Bruno, who was burnt to death as a religious heretic by the Catholic Church for maintaining that the only difference between the sun and the stars was that the stars were tremendously far away. Newton's question triggered other scientists to speculate on what makes the sun and stars shine. However, real progress in identifying the fuel source was impossible until two things were known: how big the sun was, and how much heat it was generating.

The annual heat output of the sun was first measured by the French physicist Claude Pouillet in the early nineteenth century. It was also measured by the English astronomer John Herschel, whose father, William, had in 1781 made the sensational discovery of Uranus, the first planet unknown to the ancients. John Herschel was no slouch himself. He made major contributions not only to photography – where he introduced the terms 'negative' and 'positive' – but also to astronomy. In 1834, he had sailed with two telescopes and his wife and children to the Cape of Good Hope. On what is now a hilly suburb of Cape Town, but at the time was an island surrounded by hippo-infested marshes, he set up the first major observatory in the southern hemisphere. Over the following four years, he had made the first comprehensive map of the southern sky for the purpose of improving navigation at sea. It was in 1837, the year before he had completed the mammoth task, that he found time to carry out his experiment to measure the heat output of the sun.

According to Herschel's estimate, which was in close accord with Pouillet's, enough solar heat arrived at the earth each year to melt a hypothetical blanket of ice 31 metres thick. This is a peculiar statement whose true meaning becomes apparent only once it is realised that the sun is about 150 million kilometres from the earth and radiates its heat not just in the direction of the earth but in all possible directions. In other words, the sun pumps enough heat into space each year to melt a 31-metre-thick shell of ice with a diameter of 300 million kilometres. This is enough ice to fill about 500 earths. And the sun is capable of melting this quantity year after year after year. That the sun's heat output was truly gigantic no one had seriously doubted. However, Herschel and Pouillet's great achievement was to put a precise figure on it. Armed with their numerical estimate, it was now possible to address seriously the question of what was powering the sun.

The obvious energy source was coal, the driving force behind the Industrial Revolution. By now it was widely recognised that Anaxagoras had erred on the conservative side with his guess that the sun was not much bigger than Greece. It was actually almost a million miles across – large enough to swallow a million planet earths. So how long could a sun-sized piece of coal – surely the mother of all lumps of coal – keep on burning at the rate estimated by Herschel and Pouillet? The relevant calculations were carried out in 1848 by a young German doctor called Julius Robert Mayer. He concluded that even if the sun were made of coal and the coal were burning in an atmosphere of pure oxygen it would still go out in a mere 5000 years. This was not enough even to satisfy Bishop Ussher.

Clearly, the sun was drawing on an energy source far more powerful than coal or any run-of-the-mill fuel. But what could it be?

## THE INCREDIBLE SHRINKING SUN

Mayer was in a very good position to hazard a guess because he had been thinking long and hard about the nature of heat. He had come to the conclusion that 'chemical energy' – the kind released in the burning of wood, oil or coal – was not the only type that could be converted into heat energy. Any kind of energy would do. In 1842, for instance, he used a horse-powered machine to stir a vat of paper pulp. The stirring caused the pulp to get hotter. Thus he demonstrated that 'mechanical energy' could be turned into heat. Neither mechanical energy or chemical energy was up to the task of powering the sun. However, Mayer came up with another type which he thought might do the trick: gravitational energy.

Gravitational energy is possessed by any body in a gravitational 'field'. This energy can be converted into another form if the body falls under the force of gravity – as anyone who has been hit by a falling roof slate will attest. Falling slates may not have been far from Mayer's mind in 1848 when he came up with a possible way the sun might be converting gravitational energy into heat energy. Instead of falling slates, however, Mayer imagined falling meteorites.

Meteorites are chunks of rock which orbit the sun along with the planets, and which burn up in spectacular fireballs if they ever run into the earth's atmosphere. A meteorite that became ensnared by the sun's powerful gravity would gather speed as it fell towards the solar 'surface'. When it finally hit, it would be travelling very fast indeed and its tremendous energy of motion would be converted into heat energy. It was Mayer's suggestion that large numbers of meteorites were continually bombarding the solar surface and that this explained how the sun stayed hot. This theory was also proposed in 1853 by a Scottish hydrographer called John James Waterston. Like Mayer, Waterston had realised that a chemical fuel like coal was woefully inadequate for powering the sun and that the only plausible alternative was gravity. However, Waterston had gone one better than Mayer and come up with two different ways in which the sun might convert gravitational energy into heat. One was by sucking in meteorites from the surrounding space, as Mayer had proposed; the other was by gradually shrinking.

In fact, the 'meteoritic hypothesis' and the 'contraction hypothesis' were very closely related. For if the sun were indeed shrinking, then in effect every atom of it was falling very slowly under gravity. Instead of a small amount of meteoritic matter falling a long way, a large amount of matter – in fact, the whole bulk of the sun – was falling a short way. The end result was the same, and gravitational energy was converted into heat energy. The beauty of the contraction idea was its inevitability for a hot globe of gas floating in space like the sun. Every blob of gas within such a globe would be pulled down to the centre by gravity. In the short term, it would be prevented from doing so by the pressure of the hot gas pushing up from beneath. But, as the globe lost heat to space, the temperature of the gas would fall and so too would its pressure.*

---

* This is Charles's law: the pressure of a gas goes up or down in step with its temperature.

Gravity, which never goes away, would become irresistible and the globe of gas would contract.

A slowly shrinking sun might not appear as dramatic as a sun under constant bombardment by meteorites. However, contraction was just as effective at powering the sun. According to Waterston's calculations, the solar diameter need only contract by 280 metres a year in order to generate the annual heat output measured by Herschel and Pouillet. This was little more than a ten millionth of the sun's diameter and far too small to be detectable from the earth. By contrast, if the sun was powered by infalling meteorites its diameter would have to be increasing by 10 metres a year because of the quantity of rubble accumulating on its surface.

The ideas of Waterston and Mayer were highly speculative, and so were not taken seriously by other scientists. It did not help that neither had strong scientific connections. Mayer was a physician who had earned a living as a ship's doctor on a voyage to Java; Waterston had been a teacher of cadets of the East India Company in Bombay. Mayer's failure to receive recognition for his work on heat, and the death of two of his children in 1848, led him to throw himself from a third-floor window in a bungled attempt at suicide which left him permanently crippled. Things went from bad to worse when he later spent a spell in a mental asylum. However, seven years before his death, Mayer's pioneering work was finally recognised both in Germany and in Britain, where he was honoured by the Royal Society. Waterston's life was not as tragic as Mayer's but he still had his scientific papers on heat rejected by the Royal Society. They were published only after his death, with the Society's apologies.

Waterston and Mayer's ideas about the power source of the sun were not widely publicised. And they would very probably had been forgotten had they not come to the attention of two of the greatest physicists of the nineteenth century: Hermann von Helmholtz and William Thomson.

## THE AGE OF THE EARTH

William Thomson was one of the most honoured British scientists of all time. He was elected president of the Royal Society on five consecutive occasions. He was knighted by Queen Victoria for his role in the laying of the first transatlantic telegraph cable by the 'Great Eastern' in 1866. Finally, he was made Lord Kelvin in 1882. It is by this name that he is most often remembered, principally because of the 'absolute' temperature scale, which he devised in 1848 and which is called the Kelvin scale in his honour.

Kelvin was a prolific inventor who filed patents for scores of devices, including the mirror galvanometer and the siphon recorder, which registered the feeble telegraph signals that had travelled thousands of miles over the seabed of the Atlantic. He was also an immensely productive scientist who published more than 600 scientific papers and books on subjects ranging from heat to electricity and magnetism, from the origin of the ocean tides to the age of the earth and the sun. The age of the sun, Kelvin realised, was ultimately determined by the energy source it was drawing on. This led him to ponder the nature of that source.

In 1854, two years after he began puzzling over the question, Kelvin stumbled on one of Waterston's papers on the meteoritic theory of solar energy. He embraced the idea immediately and began promoting it enthusiastically at scientific meetings. Much to his disappointment, however, observational evidence for the meteoritic theory was not forthcoming. In fact, by 1862, the deficiencies of the idea had become all too obvious to him. According to Kelvin's calculations, the total mass of meteorites needed to keep the sun burning for a period even as brief as 3000 years was equivalent to about a five thousandth of the mass of the sun. However, if such a tremendous supply of interplanetary rubble really existed, it would have to be orbiting closer to the sun than the earth, or the earth would sweep it up as it circled the sun. This would change the speed of the earth in its orbit and the length of the year. But no such effect had ever been observed.

Furthermore, its gravitational pull should either disturb the orbits of the planets Mercury and Venus or else, if the material was too near the sun to do that, alter the paths of any comets that flew close to the sun. However, careful observations of the planet Mercury showed that it behaved just as expected. So too did comets, some of which astronomers had tracked passing many times closer to the sun than the earth. Reluctantly, Kelvin concluded that there was simply not enough meteoritic material inside the earth's orbit to provide a significant source of solar energy.

With heating by meteorites ruled out, Kelvin became convinced that the sun was being powered by contraction. It was a conclusion which had been reached several years earlier by the German physicist Hermann von Helmholtz. Helmholtz, like Kelvin, was one of the most versatile scientists of the nineteenth century. An ex-doctor in the German army who invented the flashlight that ophthalmologists use to examine the retina of the eye, he switched from physiology to physics at the age of 50. Among his achievements were the first measurement of the speed of nerve impulses, and pioneering work on the workings of the ear and the physics of sound. However, Helmholtz is best known for stating, far more clearly than Mayer, the principle of 'conservation of energy' – that energy can neither be created nor destroyed, only transformed from one form into another. It was this work which led him to wonder what kind of energy the sun was transforming into heat.

Helmholtz had learnt of the contraction theory in 1853, when he had attended a talk given by Waterston at the annual meeting of the British Association for the Advancement of Science. He had been captivated by the idea. One reason was that he believed the entire solar system – the sun and its planets – had formed from a diffuse 'nebula' of interstellar gas which had shrunk under its own gravity. What could be more economical than that the same contraction process was continuing today, and was responsible for making the sun shine? One thing was clear to both Helmholtz and Kelvin. If the contraction idea were correct, the sun could not have illuminated the earth forever. Once the sun had shrunk to a pinprick, there would be no more gravitational energy available.

Kelvin imagined that the sun had shrunk to its present size from a huge and diffuse 'nebula'. In this process a tremendous amount of gravitational energy would have been converted into raw heat – 100 million times the quantity that the sun radiated into space in a year. 'It seems, therefore, on the whole probable that the sun has not illuminated the earth for 100 million years, and almost certain that he has not done so for 500 million,' concluded Kelvin. Confirmation of this figure for the age of the sun came from Kelvin's estimate of the age of the earth, made the same year. He assumed that the earth had begun its life as a ball of white-hot molten rock and had been cooling inexorably ever since. Knowing the rate at which heat was flowing outward from the earth's interior – a quantity that could be estimated from temperature readings taken down mines and deep wells – Kelvin was able to deduce how long the earth would have taken to cool from its initial molten state to its present temperature.

The answer was 200 million years. And, since the earth was uninhabitable unless bathed in constant sunshine, the sun had to be at least this old. Later, because of uncertainties in his calculations, Kelvin revised this estimate, settling on an age of only 30 million years for the sun. Thirty million years was less than 200 million years but it was still an unimaginable length of time – plenty of time for everyone. Everyone, it turned out, except the geologists.

# THE CLOCK IN THE ROCKS

In the eighteenth century, for the very first time, people had begun to estimate the age of the earth from scientific observations rather than from religious calculations. What those people – the first geologists – had discovered was overwhelming evidence that the natural processes that were moulding the planet today had been operating over vast tracts of time.

For instance, on Madeira, a volcanic island off the northwest coast of Africa, sea shells were common on the summit of the tallest mountain. The only reasonable explanation was that the summit had begun its life beneath the ocean and that immense forces within the earth had pushed it skywards until it towered more than 6000 feet above the Atlantic. Since the movement of the earth's crust was generally so slow as to be unnoticeable in a human lifetime, the first geologists concluded that Madeira's rise from beneath the waves must have taken many millions of years.

The interval of time required to elevate a volcanic island like Madeira paled into total insignificance, however, compared with the time needed to create some common types of rock like sandstones and shales. Such rocks, when inspected closely, turned out to be made of countless tiny grains fused into a single solid mass. The structure could have come about if the grains had settled out of water as a 'sediment', perhaps on the bed of an ancient lake or ocean. As more and more material trickled down from above, the layer of sediment would have become steadily thicker until eventually the most deeply buried grains would have become cemented together by the weight of the material bearing down from above.

If rocks such as sandstones and shales were indeed 'sedimentary', deposited grain by grain in this painfully slow manner, then it explained their layered, or stratified, appearance. It also explained why fossils of extinct animals were sometimes found entombed in such rocks. The corpses of animals that died in ancient seas would have sunk to the seabed, to be slowly buried by the accumulating sediment and eventually turned into stone. A similar fate would have befallen the corpses of animals which died on the land after they had been swept by streams and rivers into lakes and seas. If this explanation of the origin of sedimentary rocks was correct, then geologists possessed a crude 'clock' with which to date them. Where rock strata were exposed – for example, in cliff faces – it was necessary only to measure the total thickness of the strata. If the rate at which the sediment had been deposited in the distant past were known, it would then be possible to estimate how long the entire process had taken from start to finish.

The snag, of course, was that nobody knew how fast any sedimentary rocks had been deposited in the remote past. The rate depended on all sorts of unknown factors, such as how quickly hills and mountains were ground to dust by the action of rain and frost and the probing roots of trees. It was this rock dust, after all, swept along by streams and rivers, that was the source of much of the sediment that ended up suspended in the waters of lakes and seas. The only possible course open to geologists was to assume that ancient sediments had been deposited at the same rate that silt was being deposited in the deltas of rivers like the Nile today. The pertinent question to ask was therefore: how long would a river like the Nile take to build up a layer of mud as thick as, say, the Jura mountains in France?

When the geologists did the relevant calculations, they concluded that the earth's sedimentary rocks must have been laid down over at least 200 million years. And since sedimentary rocks, by their very nature, were chips off older blocks, the earth had to be even older than this. Assuming that the time needed to make these older rocks was comparable to the time needed to make sedimentary rocks, the age of the earth could easily be 400 million years old. Significantly, this was more than 10 times as long as Kelvin had calculated that the sun could have been shining in the sky.

The geologists were not alone in finding that vast tracts of time would have been needed for natural processes to mould the world. The same conclusion was reached independently by the biologists. In 1859, Charles Darwin published *The Origin of Species*, in which he propounded the theory of evolution by natural selection. The theory stated that in the furious struggle for scarce food resources, the creatures in any population which were best adapted to their environment invariably survived to produce more offspring than those less well adapted. This beguilingly simple principle caused populations of creatures to continually change, or 'evolve', with time. Strong evidence of such evolution could be seen in the natural world, as well as in the fossils preserved in sedimentary rocks.

But evolution, as envisaged by Darwin, was a painfully slow process. In order to evolve the fantastic diversity of creatures which inhabited the earth today from some simple common ancestor in the remote past, it was necessary for natural selection to have operated over long periods of time – hundreds of millions, if not billions, of years. On this matter, the biologists and geologists were in complete accord. However, geology was an inexact science, and the theory of evolution by natural selection the subject of fierce and often acrimonious debate. The theory of heat, on the other hand, was tried and tested in ten thousand steam engines, and it was this that Kelvin had used to estimate the ages of the earth and sun.

If the earth was really hundreds of millions, or even billions, of years old, then the geologists and biologists had to face an awkward question. What possible power source could have kept the sun burning for so long? Kelvin was adamant that the most potent source of energy imaginable – gravity – could at most have powered the sun for a mere 30 million years. 'As for the future,' he wrote in 1862, 'we may say, with certainty, that the inhabitants of the earth cannot continue to enjoy the light and heat essential to their life, for many million years longer, unless sources now unknown to us are prepared in the great storehouse of creation.'

This, as it turned out, was a prophetic let-out clause. An unknown source most certainly did exist. It was discovered by Henri Becquerel on a cloudy Sunday in 1896 and christened radioactivity by Marie Curie.

## THE DISINTEGRATING SUN

Within only a few years of Becquerel's discovery, Ernest Rutherford and Frederick Soddy had blown apart the cosy world of nineteenth-century physics, consigning to the dustbin of history the 2000-year-old idea that atoms were indestructible. In the process of radioactive decay, atoms of one kind could change spontaneously into atoms of another, unleashing a damburst of energy.

No one had the faintest idea where that energy was coming from. But, once Pierre Curie and Albert Laborde had measured the enormous energy stored in a speck of radium, it was inevitable that someone would make the connection between radioactivity and the mysterious heat source of the sun. In fact, the connection was made by two people: John Joly, an Irish geologist, and George Darwin, the astonomer son of Charles Darwin.

Darwin had emerged from the shadow of his father with a theory that the moon had been born from molten material spun off by a rapidly rotating earth. Joly had made his name by estimating the age of the earth from both the accumulation of sediment and the accumulation of sodium in seawater, settling on a figure of about 100 million years. He would also make a significant contribution to medicine by pioneering the use of radium for the treatment of cancer. In 1903, Joly and Darwin independently proposed that the sun was powered by radioactivity. If it were true, then the sun could easily have shone for far longer than 30 million years. Such a prospect was especially attractive to Darwin. By showing that there had been plenty of time for natural selection to work its magic, he would be able to refute his father's severest critics.

The potential for radioactive heating greatly impressed Joly and Darwin, who both suggested that the phenomenon might also be responsible for keeping the interior of the earth in a permanently molten state. It was an idea which received a considerable boost the following year, when Rutherford and Soddy estimated the total heat generated by all the earth's uranium- and thorium-bearing rocks, and found it to be so great that the earth might actually be warming up rather than cooling down, as Kelvin had assumed.

The earth was not warming up. Rutherford and Soddy had overestimated the effect. Nevertheless, they had shown that the heat liberated by radioactivity was more than enough to turn the earth's rocks to treacle. Today, geologists concur with Joly and Darwin. A large part of the earth's internal heat does indeed appear to come from radioactivity — especially from rocks containing uranium, thorium and potassium.

The calculations of Rutherford and Soddy demonstrated that the heat from the earth's natural radioactivity must at the very least be slowing down the cooling of the planet. This had serious implications for Kelvin's estimate of the age of the earth. For if the planet was cooling more slowly than anyone had realised, it must have taken longer than 30 million years to have plunged from a primordial white-hot state to its present-day temperature. The implications for the sun were similar. If it were powered by radioactivity, then it could have shone in the sky with undiminished brilliance for hundreds of millions, if not billions, of years. As Rutherford and Soddy declared in 1903, 'The maintenance of solar energy no longer presents any fundamental difficulty if the internal energy of the component elements is considered to be available — that is, if processes of sub-atomic change are going on.'

It turns out that such a possibility had actually been suggested four years earlier by a prominent American geologist, Thomas Chamberlain. 'Is present knowledge relative to the behaviour of matter under such extraordinary conditions as obtain in the interior of the sun sufficiently exhaustive to warrant the assertion that no unrecognised sources of heat reside there?' asked Chamberlain in 1899. 'What the internal constitution of the atoms may be is yet an open question. It is not improbable that they are complex organisations and the seats of enormous energies.' This was prescience of a very high order. After all, when Chamberlain penned these words, Curie and Laborde's measurement of the fantastic energy set free in radioactive decay was still four years in the future, as was Rutherford and Soddy's discovery that the energy release accompanied violent convulsions inside atoms. Chamberlain was, understandably, vague on details. Joly and Darwin, however, could be far more specific. They could point to a real atomic phenomenon for the heat source of the sun – the radioactive disintegration of heavy atoms into lighter ones.

In 1903, the idea of a radioactively disintegrating sun seemed eminently plausible. However, the principal reason for suggesting the idea had been to extend the age of the earth and the sun. This was necessary only if the geologists and biologists were right about the earth being immensely old. As yet, there was no definitive proof of this. However, that proof was just around the corner. It was provided not by the geologists or biologists but by the physicists. And the clock they exploited was radioactivity itself.

## THE TRUE AGE OF THE EARTH

The realisation that radioactive atoms were not only 'seats of enormous energies', but could be used as 'clocks' for dating the earth, stemmed directly from a discovery made by Rutherford in 1900. Working with Soddy in Montreal, Rutherford had noticed that the radioactivity of a gas given off by the element thorium died away in a very distinctive manner: after about a minute a sample of 'thorium emanation' retained only half its original radioactivity, after two minutes a quarter, after three minutes an eighth, and so on. Rutherford soon found that the radioactivity of other elements died away in a strikingly similar manner. The only difference was in how fast the radioactivity was lost. Each radioactive element had its own characteristic 'half-life', a term invented by Rutherford. After one half-life, a sample retained only half its original radioactivity, after two half-lives a quarter, after three half-lives an eighth, and so on.

The half-life of thorium emanation, these days known as radon, had been relatively easy to measure in the laboratory because it was close to one minute. However, the half-lives of other radioactive elements ranged from a split second to many billions of years. That of radium was 1620 years, while uranium had a half-life of 4.5 billion years. It then occurred to Rutherford that the steady disintegration of radioactive atoms might be used as a 'clock' for determining the age of rocks. As soon as a rock formed, Rutherford reasoned, its radioactive atoms started decaying. Year in year out, they disintegrated, tiny atomic explosions going off invisibly deep inside. As a rock grew old, therefore, a greater and greater fraction of its radioactive atoms succumbed to disintegration.

Measuring this fraction was impossible with the technology of the time. However, in 1905, Rutherford came up with an alternative method which could be applied to any rocks that contained a reasonable quantity of uranium. It relied on the fact that every single radioactive decay – whether it involved uranium atoms or the radium atoms which were always associated with them – vomited out an alpha particle. Since alpha particles were essentially helium atoms, with every decay more and more helium gas would collect inside the rock. The total amount that had accumulated since the uranium mineral had formed was therefore a good measure of how many atoms had decayed, which, in turn, was a measure of the age of the mineral. Uranium, because of its enormously long half-life, was an ideal clock for dating rocks that might be many hundreds of millions of years old.

Of course, this radioactive method of dating would only work if the helium atoms really did accumulate in the pores and fissures of a uranium mineral and were not lost to the air. There was some evidence that this was indeed the case since it was in a sample of one such mineral – cleveite – that William Ramsay had made the discovery of helium. Ramsay's discovery came one year before Becquerel's of radioactivity, so no one at the time appreciated the key significance of finding helium in a uranium-bearing rock. Rutherford applied his dating technique for the first time in 1906. He took two uranium minerals – fergusonite and uraninite – and heated them to drive off their helium. Then he carefully measured the amount of gas he had collected.

He had previously estimated how much helium was generated in a single year by the uranium and radium in his samples. Assuming that helium had been produced at a constant for millions of years – which was roughly, though not strictly, true – Rutherford was then able to deduce how long his samples must have been around in order to have produced the quantity of helium he collected. His conclusion was dramatic. Each was at least 500 million years old. Although the result supported the geologists, the dating method was seriously flawed because it was almost certain that some of the helium produced by radioactivity in the rocks had long ago escaped into the air. In order to provide more reliable dates, a method was needed that did not suffer from the problem of helium leakage. Rutherford had in fact proposed such a method in 1905. Rather than measuring the amount of helium, it involved measuring the amount of lead in a uranium mineral.

The method assumed that lead was the ultimate decay product of uranium. Assuming this was true, then the amount of lead in a rock compared with the amount of uranium should grow with time and so should be a measure of the age of the rock. Minerals which displayed a small ratio of lead to uranium would be young, while minerals with a high ratio would be old. The overwhelming advantage of the method was that once the lead had formed in a uranium mineral, it stayed in place. It could not simply waft away into the air like helium. The first person to put Rutherford's method into practice was an American physicist called Bertram Boltwood. He had already confirmed Rutherford's suspicion that lead was indeed the final decay product of uranium. In 1907, Boltwood measured the ratio of lead to uranium in almost a dozen rocks. The youngest, from Connecticut, turned out to be 410 million years old while the oldest, from Ceylon, was an astonishing 2.2 billion years old.

By 1929, Rutherford had concluded, from a more sophisticated method of dating, that the earth was formed at least 4 billion years ago. Later studies of the radioactivity in rocks and in meteorites would put the birth of the earth and sun about 4.6 billion years ago. No longer could there be any doubt. The age of the earth and, by implication, the age of the sun, was billions of years rather than tens of millions of years. Kelvin was wrong and the geologists and biologists were right. Ironically, it was the physicists who had supplied the definitive proof.

The reason Kelvin got it so badly wrong had in fact been pointed out by Chamberlain in 1899. 'The fascinating impressiveness of rigorous mathematical analyses, with its atmosphere of precision and elegance, should not blind us to the defects of the premises that condition the whole process,' wrote Chamberlain. 'There is perhaps no beguilement more insidious and dangerous than an elaborate and elegant mathematical process built upon unfortified premises.' In short, calculations were only as good as the assumptions on which they were based. And, according to Chamberlain, Kelvin's assumptions stank.

Chamberlain pointed out that there was not the slightest scrap of evidence that the earth had been born as a ball of white-hot molten rock. And, as far as the sun was concerned, Kelvin had no reason to believe that matter would behave as it did on earth in an environment as alien as the solar interior. It was precisely this criticism which had led Chamberlain to make his prophetic remark that there might be unrecognised sources of heat inside the sun, and that those unrecognised sources might have something to do with the constitution of atoms themselves.

When Becquerel developed his photographic plate and saw the stunning white cross against the black, overexposed background, he could never have guessed that his discovery would eventually lead to the resolution of one of the bitterest disputes of the age. But resolve it it had. Radioactivity had provided the definitive proof that the sun really was many billions of years old. And only an energy source as concentrated as radioactivity could possibly keep the sun shining for billions of years. The case for a sun powered by radioactivity now appeared very strong indeed. But was there any real evidence that the sun contained any radioactive elements? Without ever going there and bringing back a sample to examine, how would we ever know? Remarkably, there existed a way. It hinged on a discovery made a century earlier by a German optician named Josef von Fraunhoffer.

# 4

# *Deductions from a Glimmer of Starlight*

HOW WE DISCOVERED THAT THE SAME ATOMS
PRESENT ON EARTH ALSO EXIST IN THE SUN
AND STARS, BUT THAT THE DISINTEGRATION
OF ATOMS CANNOT BE POWERING THE SUN.

Never, by any means, shall we be able to study the chemical composition
or mineralogical structure of the stars.

Auguste Comte

The most remarkable discovery in all of astronomy is that the stars are
made of atoms of the same kind as those on the earth.

Richard Feynman

Josef von Fraunhoffer was from a poor Bavarian family. Apprenticed to
a firm of Munich opticians when his parents died, he was destined for
a lifetime of lens-polishing drudgery. However, in 1808, his life was
dramatically transformed by an extraordinary stroke of good fortune. As
he hunched over a workbench, furiously polishing the lens of a pair of
spectacles, the roof of his workshop uttered a sickening groan and
collapsed on top of him. When Fraunhoffer crawled from under the
mountain of masonry, white with dust from head to foot, his employers
were so astonished and relieved by the miracle of his survival that they
pumped his hand, slapped his back and rewarded him with a large gift of
money.

This remarkable piece of generosity led to many wonderful things in
Fraunhoffer's life. It gave him a firm footing on the corporate ladder and
led in 1811 to his appointment as company director at the early age of
24. More importantly, however, it gave him financial independence,

and the freedom to carry out his own optical experiments. Those experiments would earn him a permanent place in the history books.

Fraunhoffer's great ambition in optics was to perfect a special kind of glass lens – one which did not produce an image that was fringed with rainbow colours. The colour-fringing problem had been obvious almost as soon as the Dutch spectacle-maker Hans Lippershey had patented magnifying lenses. The year after, in 1608, Galileo Galilei had incorporated such lenses into a telescope, and observed mountains on the moon, a quartet of pinprick moons orbiting the planet Jupiter and countless stars too faint to be seen by the naked eye. Maddeningly, however, the images produced by the instrument shimmered with psychedelic colours.

Half a century later, the problem came to the notice of Isaac Newton. He realised that if Galileo's crude lens telescope were ever to be improved, it would first be necessary to understand the nature of light. Newton's subsequent investigations did not, it turned out, lead him to a better lens telescope. However, they did provide important insights into why the images formed by lenses were fringed by rainbow colours.

## THE UNBEARABLE WHITENESS OF SEEING

Newton gained his insights into the nature of light in 1666, the year bubonic plague drove him from his college in Cambridge to the isolated safety of his family's farm in Woolsthorpe, Lincolnshire. It was a phenomenally productive time for the 24-year-old Newton. During his year in exile, he also formulated the theory of universal gravitation and invented the mathematical tools of differential and integral calculus.

For his investigations of light, Newton purchased several triangular wedges of glass called 'prisms'. These were known to produce the same rainbow effect as lenses but had the advantage of having a far simpler shape. For centuries afterwards, Newton would be painted standing in a darkened room at Woolsthorpe, with a thin shaft of sunlight stabbing from a small circular pinhole in a shuttered window to a prism held in his hand. Emerging from the prism and illuminating the far wall of the room was a rainbow of colours: from red through orange and yellow to green, blue, indigo and violet. Newton was not the first to discover the 'spectrum' of colours produced by a prism. He was, however, the first to find that if a second prism was put in the path of the spectrum the individual colours such as red and blue were split no further. He concluded that red and blue light were pure colours whereas white light from the sun was a composite of all the colours of the rainbow smeared together. The proof was both simple and dramatic. Newton inverted the second prism and the rainbow colours promptly merged back into white light.

The clue to why a prism split white light into its constituent colours was in the shape of the rainbow patch on the wall. Instead of being circular like the hole in the shuttered window, it was a highly elongated oval, with the colours of the spectrum fanned out along its length. Newton concluded that the prism must bend, or 'refract', some colours more than others. Now it was clear why stars and planets were fringed with colours when seen through a telescope. The lenses of a telescope affected light in much the same way as a prism. They bent different colours by different amounts, separating out the constituent colours of white light. Newton concluded that a lens telescope could never produce a perfect image, a conviction which in the course of time led him to invent a revolutionary new kind of 'reflecting' telescope which concentrated light not with lenses but with a concave mirror.

Newton was mistaken in his belief that lenses were doomed always to produce coloured images. In the mid-eighteenth century, an English optician called John Dolland realised that the problem could largely be overcome by making a two-element lens, in which a converging lens made from one kind of glass was sandwiched together with a diverging lens made out of another type of glass. In such an 'achromatic' lens the spreading out of white light into its component colours by one element was cancelled out by the other. Ironically, Newton had himself shown that such a thing might be possible, by demonstrating that white light, once unravelled into coloured threads by one prism, could be twisted back together again by a second prism.

By 1758, Dolland and his son were making telescopes whose images were virtually colour-free. But though the improvement on the primitive instrument used by Galileo was dramatic, there was still room to do better. It was the task of improving on Dolland, and fabricating a better achromatic lens, that captivated Josef von Fraunhoffer.

## THE MISSING COLOURS OF SUNLIGHT

The big problem in making achromatic lenses was in ensuring that each element of the lens bent light in precisely the right way to cancel out the effect of the other. This was difficult to achieve because the light-bending power, or 'refractive index', of different glasses was not well known. If Fraunhoffer were to make a better achromatic lens than

Dolland, he would therefore need an accurate way of measuring the refractive index.

The trouble was that even the same piece of glass bent some colours more than others. To compare the light-bending power of different glasses, Fraunhoffer therefore needed a source of light of a single colour. But even a cursory glance at the spectrum of sunlight showed that each individual colour – whether red or green or yellow – was actually smeared over a wide region, rather than possessing a sharp edge which could be used as a colour standard.

But perhaps there was some other sharp feature in sunlight that might be used as a reliable colour standard. This was the hope that spurred Fraunhoffer to take a closer look at the solar spectrum in 1814. The instrument he used was called a 'spectroscope'. Sunlight, instead of passing through a pinhole before striking a prism, passed through a long thin slit in a metal plate. This created a long ribbon-like spectrum which could be scanned from end to end with a microscope.

The spectroscope was a significant improvement over the apparatus used by Newton. Fraunhoffer would be able to use it to examine the solar spectrum more closely than anyone had examined it before. He set up his apparatus in a darkened laboratory. He waited for his eyes to adapt, then peered into his microscope. Immediately he saw something peculiar. Cutting across the ribbon of rainbow colours was a single black line, as thin as a human hair.

Fraunhoffer blinked and rubbed his eyes. Surely he must be mistaken. However, when he squinted back into the microscope, the black line was still there. In fact, now his eyes were properly adjusted to the microscope, he saw the black line was not alone. There were others, and the longer he looked the more of them he could see. The entire solar spectrum was crossed by whisker-thin black lines!

He was not the first person to see the 'missing colours' of sunlight. They had actually been spotted 12 years earlier by the English metallurgist William Wallaston. However, Wallaston's spectroscope was not as good as Fraunhoffer's, and he had counted only seven dark bands in the solar spectrum. Furthermore, he had concluded that the bands were merely the 'boundaries' between colours. Content with this rather unscientific explanation, he had proceeded to forget the whole matter. Fraunhoffer was more tenacious. He discovered that the lines were present even when he used a 'diffraction grating', a polished piece of metal scored with fine parallel grooves which split sunlight into its constituent colours much like a prism. This at least proved that the lines were not produced by the material of the prism but were inherent in sunlight.

He obtained an even better spectroscope and, by 1823, had measured the positions of 574 spectral 'lines', labelling the most prominent ones with letters of the alphabet. Had Newton discovered a black line in the solar spectrum, he would have had to describe its position as 'in the middle of the yellow region' or 'on the border between blue and green'. By Fraunhoffer's day, however, it was possible to be considerably more precise because of a crucial insight into the nature of light and colours.

## THE SIGNATURE TUNE OF THE STARS

At the beginning of the eighteenth century, an English physician called Thomas Young had discovered that light was a 'wave vibration' which travelled through empty space rather like a water wave on the sea, or a sound wave through the air.* Different colours were the optical equivalent of notes of different pitch; blue light vibrates at roughly twice the pitch of red light.

A wave which vibrates rapidly is always smaller than one which vibrates sluggishly, as anyone who has watched waves on the sea will attest. An equivalent way of describing colours is therefore as light waves of different sizes. Young discovered that the size of light was fantastically small. Its 'wavelength' – a measure of the distance between successive crests of the wave – was on average only about a thousandth of a millimetre, with the wavelength of red light being about twice that of blue light. So each of Fraunhoffer's black lines corresponded not so much to a missing colour of light but to a missing wavelength of light. It was this which enabled Fraunhoffer to be precise about their position in the solar spectrum. For instance, the lines he labelled with the letters 'H' and 'K' corresponded to light at a wavelength of 0.3968 thousandths of a millimetre and 0.3933 thousandths of a millimetre, respectively. Today, astronomers know the wavelengths of more than 25,000 'Fraunhoffer lines'. They litter the spectrum of the sun like the stripes on a supermarket bar code.

Fraunhoffer had hoped to find a single sharp feature in the solar spectrum which might act as a convenient colour standard. Instead, he

* In the nonscientific world, Young is most famous for decoding the Rosetta Stone, something he did independently of the Frenchman Jean-François Champollion. He also discovered that the human eye focuses by changing its shape and proposed a theory of colour vision in which the 'retina', the surface on which light is focused, is dotted with microscopic structures that detect red, green and violet light. In a stroke, he had found an explanation for why some people were colour blind: they had a fault in one type of receptor.

had found hundreds. Any one of the most prominent Fraunhoffer lines could be used for comparing the light-bending power of different glasses. However, improving on Dollond's lens had suddenly assumed a low priority in Fraunhoffer's life. He had stumbled on something new and baffling in sunlight, and he now concentrated all his efforts on solving the mystery.

One of the first things Fraunhoffer did was point a telescope at some of the brightest stars in the night sky and examine their light with his spectroscope. To his amazement, he discerned the same spidery black lines he had seen in the spectrum of the sun. Sometimes the light of a star contained a line not present in the light of the sun, and sometimes a line was missing from the starlight. However, there could be little doubt that what Fraunhoffer was seeing was the same kind of thing as in the sun.

It had long been suspected that the stars were other suns, shrunk to mere pinpricks of light by their sheer distance from the earth. Here was the definitive proof.

The implications were enormous. If we could understand what the Fraunhoffer lines were telling us about the sun, we might gain knowledge about the stars as well. But what *were* the Fraunhoffer lines telling us about the sun? Although Fraunhoffer struggled valiantly to answer the question, he was doomed never to get to the bottom of the mystery. For, in 1826, the apprentice optician whose life had been so dramatically changed by a piece of good fortune simply ran out of luck. At the tragically early age of 39, he contracted turberculosis and died.

It would be more than three decades before the mystery of the Fraunhoffer lines was finally solved. The man who made the breakthrough was a German physicist, Gustav Kirchoff.

## THE BIRTH OF A NEW SCIENCE

The catalyst for Kirchoff's breakthrough was Robert Bunsen, a chemist at the University of Heidelberg best remembered for the Bunsen burner (even though he did not invent the device but merely perfected it). Bunsen loved experimenting with highly toxic and evil-smelling chemicals, losing an eye in the process and nearly dying of arsenic poisoning. His eagerness to spend his days in a fog of offensive vapours did little for his sex appeal, and he never married, which was a shame because, by all accounts, he was a kindly soul. The wife of the prominent German chemist Emile Fischer once said of Bunsen: 'First, I would like to wash him, and then I would like to kiss him because he is such a charming man.'

In the summer of 1859, Bunsen embarked on a programme to determine the colour of the light given off by various chemicals when they were heated to glowing point in the flame of a Bunsen burner. If each chemical turned out to have a unique 'colour signature', it would provide a powerful tool for identifying unknown substances even if only the tiniest samples were available to chemists. The difficulty was that a particular colour of light, such as yellow, was produced by many different substances. It was still possible that each substance glowed a unique shade of yellow. But if Bunsen were to test the idea, he would need an objective way of distinguishing between subtle shades of yellow, or of any other colour.

The only stategy he could think of was to look at each of his glowing chemicals through pieces of different-coloured glass. Since red glass transmitted red light, blue, blue and so on, he might be able to characterise each shade by how much of each of these colours was produced by the flame. It was as he embarked on his programme of colour comparisons that Gustav Kirchoff rolled his wheelchair over the threshold into Bunsen's laboratory and asked what he was up to.

Kirchoff was one of the great physicists of the nineteenth century. While still a student he had formulated the laws which predict the voltage and current at any point in an electrical circuit. An accident early in his career had confined him to a wheelchair. However, his handicap had not prevented him from rising to become a professor of physics at Heidelberg, where he had become friendly with Bunsen. When Bunsen explained why he was squinting at a flame through pieces of coloured glass, Kirchoff pointed out there was a far better way to compare colours of light: by using a glass prism. Even if two shades of a colour were indistinguishable to the human eye, a prism would still bend each by a slightly different amount.

Bunsen replaced the pieces of coloured glass by a precision spectroscope, and the two men took turns to peer through the microscope at the light from the flame. What they saw astonished them. The spectrum was crossed by a handful of intensely bright lines. Bunsen tried another flame-grilled chemical. Its spectrum also displayed bright lines, though in different positions. So too did another. And another.

The true magnitude of Kirchoff and Bunsen's discovery was apparent to the two men only after several hours of feverish activity. For it was only then that they were able to confirm a growing suspicion. The pattern of bright lines was unique to each substance. Nature, very conveniently, had endowed each material with its own spectral 'fingerprint'. It was a discovery that would revolutionise the science of chemistry. Even if a substance existed in microscopically small amounts, it could nevertheless be identified by its characteristic pattern of bright spectral lines.

Within months of their discovery, Kirchoff and Bunsen had demonstrated the potential of the technique in the most dramatic way. In the light of their laboratory flame, they discovered the spectral fingerprint of an element entirely unknown to science. Its most prominent line was in the blue region of the spectrum so they christened it caesium, from the Latin for 'sky blue'. The following year they discovered yet another new element, this time with a prominent line in the red part of the spectrum. They named it rubidium.

But although Kirchoff and Bunsen's discovery had an enormous effect on chemistry, it was destined to have an even greater impact on astronomy. Here the credit goes entirely to Kirchoff. For it was he who thought to compare the flame spectra with the spectrum of the sun. Kirchoff was immediately struck by something odd. It concerned a pair of prominent dark lines in the yellow region of the solar spectrum. Fraunhoffer had labelled them with the letter 'D'. What was odd about the dark lines was that they occupied exactly the same position as a pair of bright lines in the spectrum of a flame impregnated with the element sodium, an element commonly found in table salt.

The curious match had been noticed by others before and passed off as a mere coincidence. However, Kirchoff was convinced that there was more to it. Bizarre as it seemed, there must be a connection between sunlight and sodium. In an attempt to find what it was, Kirchoff passed a beam of sunshine through a sodium flame. In the sunlight emerging from the flame, the bright lines from the sodium would be exactly superimposed on the dark D lines from the sun. The reasonable expectation was that the D lines would become lighter. However, when Kirchoff examined the light with his spectroscope, this was not what he saw at all. Incredibly, the solar D lines had actually got darker.

The only reasonable explanation for this darkening was that the sodium in the flame was absorbing sunlight at precisely the same wavelengths as the solar D lines. This had profound implications for the origin of the Fraunhoffer D lines because most of the light at these two wavelengths had gone missing before the sunlight even encountered the sodium flame. All the sodium in the laboratory was doing was merely enhancing an effect that was already there. This led Kirchoff to a startling conclusion. Somewhere, on its long journey through space to the earth, sunlight was being absorbed by another source of sodium.

In fact, it was possible to be more specific about the source of absorbing sodium because of an insight gleaned four years earlier by a Swedish scientist, Anders Angström. Angström had found that a gas always absorbs light at the same wavelength that it emits light. Furthermore, if the gas is hotter than the light source, then more light is emitted by the gas than absorbed, creating a bright line in the spectrum of the light source. If the gas is cooler than the light source, on the other hand, the opposite happens: more light is absorbed by the gas than is emitted, creating a dark line.

The dark solar D lines was therefore telling Kirchoff that the sodium gas through which sunlight was streaming on its way to the earth was cooler than the sun. But where was the sodium? There was no evidence of any sodium gas floating in space, or even in the earth's atmosphere. The only alternative Kirchoff could think of was that the sodium gas responsible for the dark Fraunhoffer D lines was floating in the relatively cool outer atmosphere of the sun.

Remarkably, there was a way to test the idea in the laboratory. Kirchoff created an artificial sun by burning a piece of chalk in a searing-hot oxygen–hydrogen torch. He passed this intensely bright 'limelight' through a cooler sodium flame, and examined the light that emerged with a spectroscope. There, crossing the spectrum of the artificial light, were two black lines at precisely the same wavelengths that a sodium flame emitted light. Kirchoff had created artificial D lines exactly like those in the sun.

Now there could be no doubt. The sun's cool atmosphere really did contain sodium. The extraordinary thing was that, without ever leaving the confines of his laboratory, Kirchoff had proved that an element present on the earth was also present in the sun. Kirchoff began scanning the solar spectrum for the characteristic spectral fingerprints of other elements. Very soon, he stopped at the dark solar lines which Fraunhoffer had labelled with the letter 'E'. It took only a moment to be sure. They exactly matched the bright lines produced when iron was heated to incandescence in a laboratory flame. He had found that the Sun contained iron as well as sodium.

The mystery of the Fraunhoffer lines was at long last solved. They were the combined 'fingerprints' of all the different atoms that existed in the sun. In time, Kirchoff was able to identify dozens of elements in the solar spectrum. Among them were calcium and carbon, silver and silicon, zirconium and zinc. The conclusion was inescapable. The sun was made of exactly the same kind of atoms as the earth.

Merely by comparing the dark lines in the solar spectrum with the bright lines produced by substances in the laboratory, Kirchoff had been able to identify the elements that made up a celestial body millions of miles away across space. It was the birth of an entirely new science – astrophysics.

In theory, astrophysics could reveal the composition not only of the sun but of the stars as well. In practice, however, the light of only a few stars – the bright ones observed by Fraunhoffer himself – was strong enough to be examined with a spectroscope. What changed everything was the advent of the photographic plate, with its ability to soak up far more feeble starlight than can the human eye.

In England, a wealthy amateur astronomer called William Huggins spent a lifetime recording the spectra of stars. By the time he was 84 years old and too blind to continue, he had recorded the spectra of hundreds upon hundreds. And, in spectrum after spectrum, he saw the unmistakable fingerprint of familiar elements – from iodine to iron, silicon to sodium, magnesium to manganese. Once again, the conclusion was inescapable: the stars were made of exactly the same kind of atoms as the earth.

It was a remarkable result, deduced from the merest glimmer of starlight. And it was all the more remarkable for having been ruled out as totally impossible only a few decades earlier by the French philosopher Auguste Comte. In 1835, Comte made a categorical statement about what we might and might not know about the stars. Although he could see no fundamental reason why we might not one day be able to measure the distance of stars, their movements, sizes and shapes, he deemed it self-evident that we would never be able to study their chemical composition.

In 1835, when Comte made this statement, it was a safe bet. The stars were generally thought to be enormously far away, so it was unlikely that anyone would ever travel to one. And unless someone did, and brought a sample back, how would we ever know what they were made of? Comte's statement was perfectly reasonable. Reasonable but wrong. Thanks to Kirchoff, within only two years of Comte's death in 1857 science was making huge strides in determining the composition of stars. Nature, it turned out, was not nearly as unkind to us as Comte had supposed.

But the new science of astrophysics, though it provided breathtaking new insights, also threw up new and baffling puzzles. One came to light within a decade of Kirchoff's breakthrough: a prominent spectral line in the sun which corresponded to no known element. It was discovered by a clerk at the British War Office called Norman Lockyer.

## AN ATOM UNIQUE TO THE SUN

Norman Lockyer was a remarkable man. Although left to bring up seven children when his first wife died, he nevertheless found the time to write the first book on the St Andrew's rules of golf, found London's Science Museum in South Kensington, and launch the prestigious international science journal *Nature*, which he ran as Editor for 50 years. But it was as an astronomer, first amateur, later professional, that Lockyer distinguished himself.

In 1862, at the age of 26, he built a six-inch telescope in his garden in the south London suburb of Wimbledon. At first, he used it simply to observe Mars and the other planets. But now that the mystery of the Fraunhoffer lines had been solved by Kirchoff the lure of the sun was irresistible. Lockyer bought himself a spectroscope with a view to studying the message written in sunlight.

Something that particularly interested him was the problem of observing solar 'prominences', filaments of gas hundreds of thousands of kilometres long, which were periodically coughed out by the sun and which could surge far into space before looping back down to the surface. Prominences were cooler and fainter than the sun, which made them impossible to see in normal circumstances. Only during a total eclipse, when the moon blotted out the blinding glare of the solar surface, were prominences visible – rose-coloured arches arrayed around the lunar circumference. Unfortunately, total eclipses were rare at any given place on earth, and lasted at most a few minutes. What astronomers needed was a means of observing prominences at times other than total eclipses. It was just such a method that Lockyer dreamt up in 1866.

Prominences, he reasoned, would produce bright spectral lines because they had only the cold of space behind them, rather than the intense heat of the sun. Such lines should be detectable at all times – even when there was no total eclipse. All Lockyer needed to see them, and to find out what the gas of solar prominences was made of, was a better spectroscope. Because it took some time to have such an instrument built, Lockyer could not test his idea immediately. However, on 20 October 1868, he pointed his telescope at the edge of the sun and examined the light. To his delight, he saw bright spectral lines from a solar prominence. As it happened, the very same lines had been observed from India two months earlier, by the French astronomer Pierre Jules César Janssen. He too had had the idea that prominences might be detactable at times other than total eclipses. Among the bright lines from a solar prominence, Lockyer and Janssen both noticed a mysterious line which neither had seen before. It was in the yellow region of the spectrum, and very close to the wavelength of the twin D lines given out by glowing sodium.

In their respective laboratories both men heated up large numbers of gases in an attempt to find one that produced the mysterious yellow line. But though they searched long and hard they did not succeed. The line did not appear to correspond to any known element. By 1870, Lockyer was confident that he had exhausted all the possibilities. He proposed that the unknown line was the fingerprint of an element hitherto unknown on earth.

Lockyer gave the element the name 'helium', from the Greek word 'helios', meaning sun. It was the very same gas that William Ramsay would stumble on a quarter of a century later in the uranium–mineral cleveite. In fact, Ramsay would find it by its characteristic yellow spectral line, a dramatic demonstration of the ability of spectroscopy to reveal tiny quantities of a substance.

Helium has the distinction of being the only element to be discovered on the sun before it was discovered on the earth.* It had not been found earlier because it is both inert and lighter than air. Its inertness means that it almost never becomes trapped in compounds with other elements, while its extreme lightness means that as soon as it is released into the air it floats off into space.

Helium on earth is intimately connected with radioactivity. The association was established by Ramsay and Frederick Soddy in 1903, and later explained by Ernest Rutherford and Hans Geiger when they showed that the alpha particles spat out by radioactive elements were electrically charged helium atoms. The reason that helium was associated with radioactivity was simply that it was created by radioactive atoms. It was no accident that Ramsay had found helium trapped in a rock containing a radioactive element like uranium. But if helium on earth came from radioactive atoms, then what about the helium Lockyer had observed in solar prominences? Rutherford had not the slightest doubt. It too must come from radioactive atoms. The sun must be radioactive! In 1904, he pointed out that the presence of helium in the sun was powerful evidence for George Darwin and John Joly's idea that radioactivity was the ultimate source of sunlight.

However, although the evidence was persuasive, the fact remained that helium was merely a byproduct of radioactivity. The definitive proof that radioactivity powered the sun would of course be the

---

* The irony is that helium is the second most common element in the entire universe, after hydrogen.

observation of the radioactive atoms themselves. It was here, unfortunately, that Joly and Darwin's idea received a fatal blow. Although there was abundant evidence in the sun's spectrum of stable elements such as iron, calcium and copper, nowhere could astronomers find Fraunhoffer lines caused by radium. Nor was there any sign that uranium and thorium were present in the sun in anything but minuscule quantities.* Whatever the source of solar helium, it was certainly not radioactive decay.

It was a conclusion which left physicists in a bind. Radioactivity was the only power source that could conceivably keep the sun hot for the billions of years the earth had existed. However, there appeared to be no radioactive elements on the sun, so radioactivity could not possibly be responsible for sunlight. One thing, however, was still undeniable: the prodigious energy locked up inside atoms was easily sufficient to heat the sun for billions of years. The question therefore arose: could there be another way of releasing atomic energy besides radioactivity?

It turned out that there was. It was discovered by a Cambridge physicist called Francis Aston.

---

* In fact, even if the sun were made entirely of uranium, it would still shine only half as bright as the star we orbit. The very long half-life of the element ensures that the energy locked inside uranium atoms comes out at a trickle rather than a flood as in radium.

# 5

# The Anomalously High Mass
# of Hydrogen

HOW WE DISCOVERED THAT THE CONVERSION
OF A LIGHT ATOM INTO A HEAVIER ATOM
WOULD UNLEASH A DAMBURST OF ENERGY,
AND HOW THE SUGGESTION WAS MADE
THAT SUCH A PROCESS WAS
POWERING THE SUN.

The smallest particles may cohere by the strongest attractions.

Isaac Newton

$E = mc^2$? Very good, Albert, but show your working.

Graffiti, Cambridge

Francis Aston was the son of a wealthy merchant from the Midlands. He used his inheritance to indulge his passions for travel and sport, taking trips around the world, skiing and racing motor cars. On a visit to Hawaii in 1909, he even learnt to surf off Honolulu's Waikiki beach. Ultimately, however, the idle life of a rich playboy was not for Aston. A shy man who never married, he instead dedicated his life to scientific research.

The course of Aston's life was set irrevocably when he became J. J. Thomson's assistant at Cambridge in 1910. At that time the discoverer of the electron was becoming increasingly interested in the problem of positive or canal rays.

Canal rays had been discovered in 1886 by the German physicist Eugen Goldstein.[*] While using a discharge tube with a channel, or

---

[*] It was Goldstein, incidentally, who gave cathode rays their name.

'canal', drilled through its cathode, Goldstein had noticed that the rarefied air on the side of the cathode facing away from the anode glowed with a soft light. He suggested that the atoms of the thin air were being struck by some kind of invisible rays. Since the rays were stabbing through the hole in the cathode, they had to be coming from the positive electrode, or anode. The crucial clue to the nature of canal rays came when scientists tried to deflect them with electric and magnetic force fields in much the same way that J. J. Thomson had deflected cathode rays. Canal rays turned out to be much harder to nudge from their course, implying they were made of particles thousands of times heavier than electrons. This made their masses suggestively close to those of atoms.

If the canal ray particles were atoms, then they had to be atoms without some of their electrons, since electric and magnetic fields deflected them in the opposite direction to negative cathode rays, indicating that they carried a positive charge. The best explanation was that canal rays were atoms of the thin gas which had been struck violently by cathode ray particles careering towards the anode. The collisions knocked out some of their electrons, leaving behind positively charged atoms, or 'ions', which were then attracted to the negatively charged cathode. Although most of these ions slammed into the metal of the electrode, a small fraction passed right through the hole. It was these, emerging on the other side, which made up canal rays.

Solving the puzzle of canal rays shed light on what was actually going on inside a discharge tube when electricity passed between its electrodes. Hurtling in one direction along the tube were cathode ray electrons, occasionally colliding with gas atoms. Drifting in the opposite direction – much more sluggishly because of their larger mass – were positive gas atoms stripped of an electron or two in the collisions. But the significance of canal rays went far beyond simply illuminating the esoteric processes going on inside a discharge tube. As J. J. Thomson was the first to recognise, canal rays offered a powerful means of weighing different atoms. Just as the deflection of cathode rays in electric and magnetic fields had revealed to Thomson the mass of their constituent electrons, the deflection of canal rays might yield the mass of their constituent atoms.

What kind of atoms they were depended only on the thin gas in the discharge tube. If it was nitrogen, then the canal rays would consist of a thread-thin stream of electrically charged nitrogen atoms; if it was helium, then the rays would consist of a beam of charged helium atoms. Using the deflection technique, Thomson was able to make moderately accurate estimates of the masses of a range of different atoms. The work was important, though unexciting. However, in 1913, it threw up something unexpected. Instead of a single glowing spot where the beam of atoms struck the wall of the glass tube, there were two.

The gas responsible for the peculiar effect was neon. One possibility was that it came in two distinct atomic forms; if each form had a different mass, it would naturally be deflected by a slightly different amount. But this was just a guess. Thomson needed proof. The man he turned to supply that proof was his motor-racing, surfing assistant Francis Aston. The technique that Aston brought to bear on the neon problem was laborious in the extreme. It involved passing the gas through porous pipe clay. In theory, lighter atoms should 'diffuse' through such a material more easily than heavier atoms. So, if neon really did consist of two different types of atom, the gas that emerged from the clay should be very slightly enriched with the heavier atoms. After many passages through the clay, the gas ought to be measurably lighter than normal.

This was exactly what Aston discovered. He concluded that neon consisted of two distinct isotopes. The lighter one, neon-20, weighed 20 times as much as hydrogen and made up 90 per cent of the gas; the heavier one, neon-22, weighed 22 times as much as hydrogen and accounted for the remainder of the gas. Now it was clear why the atomic weight of neon had been found to be 20.2. The figure was simply the average weight of its different atoms.

As neon is a non-radioactive element, the discovery that it consisted of a mixture of isotopes demonstrated that isotopes were not simply a peculiarity of unstable elements like uranium and radium. They were a general feature of nature. But the real significance of the neon discovery was that it convinced Aston of the enormous potential of the deflection technique for weighing atoms. Thomson had achieved his success with relatively crude apparatus and Aston was sure he could improve on it. Unfortunately, the outbreak of the First World War put an end to pure scientific research, and for the next four years Aston was obliged to work on the development of tougher fabrics for aircraft coverings. He nevertheless kept alive the dream of building a precision instrument for weighing atoms and, in 1918, when he was finally able to return to his Cambridge laboratory, he began its construction.

## THE VINDICATION OF WILLIAM PROUT

Aston's instrument, which was called a 'mass spectrograph', was built around a discharge tube. Canal rays travelling through the tube passed through narrow slits in two successive metal plates, creating a knife-sharp beam of atomic nuclei. This was then split into several beams by an electric field, which deflected each type of nucleus through a slightly different angle. Finally, a magnetic field bent each beam by an amount that depended on the mass of its particles.

Whereas Thomson had measured the deflection of his atomic beams simply by noting the position of fuzzy blobs of light on the wall of his discharge tube, Aston employed a more sophisticated method. He placed a sensitive photographic plate in the path of his atomic beams. This not only enabled him to record the positions of the beams more accurately than Thomson; it also provided him with a permanent record.

Once his mass spectrograph was perfected, Aston was able to weigh the atoms of more than 50 different elements to an accuracy of 1 part in 100,000. By comparison, an accuracy of only about 1 part in 10 had been enough to distinguish between neon-20 and neon-22. The first discovery Aston made with his mass spectrograph was that William Prout's rough rule of thumb was a lot more than a rough rule of thumb. In 1815, Prout had pointed out that the masses of atoms were close to whole-number multiples of the mass of hydrogen, the lightest atom. Prout had concluded from this that all elements must be assembled from a basic building block, which he quite reasonably assumed was the hydrogen atom. His conclusion had been undermined, however, by a handful of elements whose weights were not at all close to whole numbers, most notably chlorine, 35.5 times as heavy as hydrogen.

One possible explanation for this departure from Prout's rule was that rogue elements like chlorine were mixtures of atoms with different masses – mixtures of isotopes in the modern terminology. This possibility had been suggested in 1871 by the English physicist and spiritualist William Crookes and for more than four decades had hung in the air, impossible to test. The advent of the mass spectrograph, however, changed everything.

Aston quickly discovered that chlorine came in two stable forms. The lighter isotope, chlorine-35, made up 77.5 per cent of the gas, while the heavier one, chlorine-37, accounted for the remainder of the gas. This made the average atomic weight of the element 35.5 times that of hydrogen – exactly the value it had been given.

The mass spectrograph revealed that other rogue elements were also mixtures of isotopes. Crookes was vindicated and so too was Prout. His rough rule had proved to be universally binding. It applied to every isotope of every element.

The evidence that atoms were built from some fundamental building block was now very strong. Prout had assumed that the block was the hydrogen atom – in modern eyes, the proton, or hydrogen nucleus. However, strictly speaking, all Prout's rule indicated was that the nuclear building block had the same mass as the proton. In Prout's day, and even when Aston began using his mass spectrograph in 1919, the only particle with the mass of a proton was the proton itself. Nobody, not even the great Ernest Rutherford, had guessed that nature concealed another, equally massive, particle, and that the atomic nucleus was built not from one building block but from two: the proton and the neutron.

Aston continued to measure atomic masses throughout 1919; then, just when he least suspected it, his mass spectrograph turned up something odd: a definite departure from Prout's rule. The departure was extremely small but it would turn out to have enormous significance for science. It would shed light on the mysterious source of atomic energy and provide the strongest hint yet of what was driving the sun and stars.

## THE ANOMALOUS MASS OF HYDROGEN

Strictly speaking, Aston's mass spectrograph did not weigh atoms. It was impossible to say, for instance, that an atom weighed so many billionths or trillionths of a gram. What the instrument in fact did was compare the weights of different atoms.

When such comparisons are made, it makes sense to choose a standard against which every measurement is compared. Aston, fairly arbitrarily, selected the mass of the oxygen atom. Rather than giving it a mass of 1, thereby endowing all lighter elements with a fractional mass, Aston instead gave oxygen a mass of 16. Today, carbon not oxygen is used as the atomic mass standard, and it is assigned a weight of 12. The masses that follow are therefore those which Aston would have obtained had he compared all atoms with carbon rather than oxygen.

Aston's measurements had shown that Prout's rule applied to a degree of precision Prout could never have dreamt of. The masses of most atoms were more than simply close to whole numbers, they were indistinguishable from whole numbers. There was one glaring exception: hydrogen. If Prout's rule was exact, the mass of the hydrogen atom should be 1. Instead, Aston measured it to be 1.008.

The hydrogen atom was 0.8 per cent heavier than it had any right to be.

The implications for Prout's hypothesis were very serious. If hydrogen was indeed the building block of all atoms, as Prout had claimed and Aston's measurements had appeared to confirm, it was reasonable to expect an atom made of four hydrogen atoms would weigh four times as much as hydrogen, an atom composed of 12 hydrogens 12 times as much, and so on. The anomalously high mass of hydrogen, however, meant that this reasonable expectation was never fulfilled.

If a helium atom was made from four hydrogen atoms, its mass ought to be $4 \times 1.08 = 4.036$. However, Aston's measurements showed that it was 4. Similarly, if an oxygen atom was made from sixteen hydrogen atoms, its mass should be $16 \times 1.008 = 16.128$. In reality, it was 16. And helium and oxygen were not unique. If hydrogen weighed 1.008, then the weight of every atomic nucleus was less than the weight of its constituent hydrogen atoms.

It defied all reason. When two 1-kilogram masses were placed on a set of scales, they always registered 2 kilograms; they never registered 0.8 per cent less than 2 kilograms. It seemed that in the world of atoms mass was capable of vanishing into thin air.

As Aston puzzled over his discovery, he recalled a daring speculation about atoms made more than half a century earlier by a reclusive French chemist, Jean Charles Gallisard de Marignac. De Marignac had spent 30 years in a cold, damp cellar at the University of Geneva, painstakingly measuring the atomic masses of the common elements. The work had wrecked his health. However, it had yielded mass estimates which, though admittedly not in Aston's league, were excellent for their time.

Like many other scientists, de Marignac had puzzled long and hard over rogue elements like chlorine, whose masses were not simple multiples of the mass of hydrogen. However, he did not conclude, as did Crookes, that such elements were mixtures of different atoms, each of which individually obeyed Prout's rule. De Marignac's alternative view, published in 1861, was that every element violated Prout's rule to a lesser or greater extent. The only thing special about an element like chlorine was that the violation was big enough to be noticeable. As to why Prout's rule should be inexact, de Marignac came up with a startling explanation. When hydrogen atoms come together, or 'condense', to form other elements, he suggested, mass is lost as energy.

Mass is lost as energy.

Sixty years later, de Marignac's words resonated in Aston's mind. His measurements with the mass spectrograph had convinced him that hydrogen was indeed the basic building block of all atoms. However, they had also led him to the absurd conclusion that each atom weighed less than the sum of its parts. The only way Aston could see to reconcile these two apparently irreconcilable observations was to assume, as de

Marignac had, that when hydrogen atoms coalesced to make other elements, mass disappeared as energy.

In 1861, such a suggestion was outrageous. By 1919, however, it was an idea which even a respectable scientist like Aston might entertain. What had changed everything was a revolutionary theory of space and time, published in 1905.

## THE ILLUSION OF SPACE AND TIME

Albert Einstein's special theory of relativity had resolved a highly embarrassing conflict between two of the great theories of physics. The conflict had centred on the speed of light.

According to the theory of 'electromagnetism', developed by the Scottish physicist James Clerk Maxwell in the 1860s, light was an 'electromagnetic' wave. Since nobody had suspected a connection between optics on one hand and electricity and magnetism on the other, Maxwell's theory was a great surprise. But there was an even bigger surprise in store. The theory predicted that when electromagnetic waves travelled through empty space, they always travelled at 300,000 kilometres per second — a million times faster than sound waves. However, the really astonishing thing was not the size of the speed of light but the fact that it was always the same. For Maxwell's theory maintained that the velocity of light was unaffected by either the motion of the source of the light or the motion of whoever was observing the light.

Immediately, this brought Maxwell's theory of electromagnetism into conflict with 'mechanics', the theory of motion as formulated by Galileo and Newton. According to the laws of mechanics, if a source of light were to approach someone at, say, 150,000 km/s, the 'observer' ought to measure the light arriving at a speed of 300,000 km/s + 150,000 km/s = 450,000 km/s. Similarly, if an observer were to travel towards a light source at 150,000 km/s, they ought to measure the light arriving at 300,000 km/s + 150,000 km/s = 450,000 km/s. It was simply the common sense that tells two car drivers racing towards each other at 100 km/hr that they are destined to collide at a combined speed of 200 km/hr.

But Maxwell's theory, flying in the face of this simple intuition, predicted that regardless of whether an observer was approaching a light source at 150,000 km/s or the light source was moving towards an observer at 150,000 km/s the observer would always measure the light arriving at 300,000 km/s.

Electromagnetism and mechanics could not both be correct. Only Einstein, however, had the bare-faced audacity to question mechanics, a theory which had never, in more than two centuries, been found wanting.* Einstein's belief that mechanics was flawed had in fact been long in gestation. Ever since his late teens, he had been aware that the theory was founded on dubious assumptions. One was that there was such a thing as absolute speed – that is, that the speed of a body was something unique which everyone observing the body would agree on. Hard thinking on this matter, however, had convinced Einstein that this was not the case.

Newton's recipe for measuring the speed of a body moving through space involved simply timing it as it passed between two fixed points. However, Einstein realised that this procedure was founded on two rickety assumptions.

The first was that time flowed at the same rate for everyone – that is, there was such a thing as a universal, or 'absolute', time. Only if this was the case would two observers be able to agree on how long it took a body to pass between two points in space.

The standard way for two observers to check that their clocks were telling the same time was to compare them, adjusting them so that they remained synchronised. This required the observers to continually inform each other of the time on their respective clocks. Newton, by claiming there was such a thing as universal time, was tacitly assuming that such timing information could be exchanged instantaneously – that is, at an infinite speed.

But, according to Maxwell's theory, this was impossible. His insistence that the velocity of light appeared the same to everyone irrespective of their speed was the same as saying that you could never catch up light. Even if a person could chase a receding beam at 99 per cent of the speed of light, they would still find it barrelling way from them at the speed of light. Since light was terminally out of reach, its velocity must represent the ultimate cosmic speed limit. Consequently, if two observers wished to synchronise their clocks the best they could do was exchange timing information at the speed of light. And light, though stupendously fast, was not infinitely fast.

The effect on the traditional concept of time was devastating. According to Einstein, clocks must run at different rates depending on

---

* In fact, the constancy of the speed of light – in blatant defiance of the laws of mechanics – had been established experimentally by two Americans, Albert Michelson and Edward Morley, in 1887. However, the dramatic result of the Michelson–Morley experiment was unknown to Einstein when he began his attempts to reconcile mechanics and electromagnetism.

how fast they were moving relative to each other. And what was more, the discrepancy between two observers' clocks became increasingly dramatic as their relative speed approached that of light.

The relativity of time had been completely hidden from humanity because the speed of light was so much greater than anything else in normal human experience. Compared with light, all familiar objects – from cars to trains to supersonic planes – crawl along in ultra-slow motion.

But time was not the only casualty of the finite and constant speed of light. The impact on space was every bit as shattering. For the second shaky assumption on which Newton had based his recipe for measuring speed was that two observers would always agree on the distance between any two points in space. Certainly, if space was a thing, like an artist's canvas on which two points could sit like two blobs of paint, there was no doubt that two observers would always agree on the separation of two points. You could imagine cutting out the strip of space between the points and measuring its length with a ruler just like cutting and measuring the strip of canvas between two blobs of paint.

But was space really a cosmic canvas that existed independently of the material bodies in the universe? Newton believed it was. But Einstein saw no evidence for such 'absolute space'. To him, space was merely the chasm between material bodies. No more, no less. Without material objects to fill the universe there would be no space.

Measuring the separation between two bodies in empty space, just like synchronising two clocks, could be done no faster than light. Einstein imagined a scheme in which an observer bounced light beams off the bodies and measured their separation from the time taken for the light beams to return and the speed of light. The consequences of measuring space using light, with its finite and constant speed, turned out to be as profound as they had been for measuring time. According to Einstein's calculations, observers moving relative to each other would measure a different distance between two bodies in space. One person's space was not the same as another person's. And, as with time, the discrepancy became increasingly dramatic as their relative speed approached that of light.

Einstein's verdict was unequivocal. There was no absolute space against which the position of bodies could be measured and there was no absolute time against which all clocks could be compared. Both space and time were relative; Newtonian mechanics was built on shifting sand. In its place Einstein founded a new mechanics – 'special relativity' – on the only thing that remained solid: the rock-like constancy of the speed of light.

The theory of special relativity, published the same year as Einstein's atomic theory of Brownian motion, predicted that a moving clock ran more slowly than a 'stationary' clock – that is, from the viewpoint of one observer moving relative to another, a body took less time to travel between two points. Of course, if the 'body' travelling between the two points was a light beam, both observers would have to agree on the velocity since the speed of light must be the same for everyone. The implication was that a moving observer, in addition to measuring a shorter time of transit, must also measure a shorter distance between the two points so that the two exactly compensated for each other.

Here then was the resolution of the conflict between electromagnetism and mechanics. For an observer in motion, time 'dilated' and space 'shrank'. And the dilation and the shrinkage, like some incredible cosmic conspiracy, always happened in such a way as to make the speed of light appear precisely the same to everyone.

But space and time did more than merely stretch like elastic for a moving observer. Space and time owed their very existence to the web of light criss-crossing the universe. What appeared to one observer as an interval of space might appear to another observer moving relative to the first as an interval of space and time. And what appeared to one observer as an interval of time might appear to another observer moving relative to the first as an interval of time and space.

Space and time were the same thing. Our senses were simply hoodwinked into believing they had a separate existence because we lived in a slow-motion world. If we lived in a world where everything moved close to the speed of light, there would be one seamless entity: spacetime. But whereas space and time were relative, spacetime was absolute, a truth recognised by Einstein's ex-mathematics professor Hermann Minkowski in 1908.* It could be warped by massive objects – as Einstein would show in 1915 – and it could come to an abrupt end inside black holes – as the English physicist Roger Penrose would show in 1964. However, spacetime, not absolute space, was the canvas on which the cosmic drama was played out.

Einstein's 'unification' of space and time had tremendous ramifications; the entire edifice of physics had hitherto been built on the twin foundation stones of space and time. The fact that one person's space was another person's time, for instance, led to the realisation that one person's electric field was another person's magnetic field. Again, we had believed that electric and magnetic fields had a separate existence because of the slowness of our world. In reality, there was only one seamless entity: the electromagnetic field.

---

* Minkowski had believed his student to be a 'lazy dog' who would amount to nothing.

But of all the unifications brought about by Einstein's special theory of relativity, without doubt the most startling was the unification between mass and energy.

## THE EQUIVALENCE OF MASS AND ENERGY

The link between mass and energy could also be seen as a direct consequence of the unreachability of the speed of light. Common sense said that pushing a massive body harder and harder would cause it to move faster and faster until eventually it reached, then exceeded, the speed of light. Since this was impossible, a body must offer some kind of resistance to being pushed and this resistance must become enormous close to the velocity of light, so that the ultimate speed was never attained.

The only conceivable source of such resistance was a body's mass. Somehow, it must get heavier and heavier as it got faster and faster. In other words, the body's increasing energy of motion must manifest itself as increasing mass.

A peculiar consequence of such reasoning is that a material body must get heavier as it gets hotter because its constituent atoms are moving about faster. It is an extremely small effect. Nevertheless, a litre of water at 100°C weighs a few million millionths of a gram more than it does at 0°C.

But how can energy of motion be changed into mass? A fundamental tenet of physics is that energy can neither be created nor destroyed, only metamorphosed from one form into another. The only conceivable answer, Einstein concluded, was that mass itself is a form of energy.

The consequences of this discovery were twofold. Not only was it possible to convert other forms of energy – for example, energy of motion – into mass-energy, it was also possible to transform mass-energy into other forms of energy – for instance, heat energy. The formula Einstein discovered for the quantity of energy locked up in a chunk of matter is perhaps the most famous, and least understood, formula in all of science:

$$E = mc^2$$

(where E represents energy, m the mass of a body and c the speed of light)

Since the speed of light is a very large number, and it enters into

Einstein's formula multiplied by itself, it follows that even a small amount of matter contains a huge amount of energy. In fact, mass-energy is so concentrated that a single kilogram of matter could in theory liberate the same amount of heat as 100,000 tonnes of dynamite.

The equivalence between mass and energy has far-reaching implications because it is universal. All energy, whether it be electrical energy, sound energy or chemical energy, has an equivalent mass. And just as the addition of energy to a body of matter increases its mass, the loss of energy by a body of matter reduces its mass. This means that a chemical fuel such as coal loses mass when it gives out heat. In fact, the process of burning is hugely inefficient, and a mere hundred millionth of the mass of a piece of coal is turned into heat energy. In other words, when a 1-kilogram lump of coal burns thoroughly, the total weight of the ash plus gases created is a hundred millionth of a kilogram less than the weight of the initial lump. Such a mass deficit is far too small to measure, ruling out a simple verification of the equivalence of mass and energy.

But Einstein was not a man to be easily put off. In the 1905 paper in which he announced to the world the dramatic and unexpected connection between mass and energy, he speculated: 'It is not impossible that with bodies whose energy content is variable to a high degree the theory might be put to the test.'

By bodies with highly variable energy content Einstein meant radioactive bodies. Only two years earlier Pierre Curie and Albert Laborde had found that radium generated about a million times as much energy as an equivalent mass of a chemical fuel like coal. It was clear that this loss of energy translated into a substantial loss of mass. Whereas the burning of coal turned a mere hundred millionth of the coal's total mass-energy into heat energy, the decay of radium must involve the destruction of close to a hundredth of its mass.

It was a substantial mass loss. However, detecting it was still far beyond the capabilities of experimenters in 1905, since the biggest sample of radium they could lay their hands on amounted to little more than a speck.

## THE ULTIMATE SOURCE OF ATOMIC ENERGY

The real significance of Einstein's speculation about radioactive substances was that at long last someone had identified the source of the tremendous energy unleashed by disintegrating atoms. Regardless of the details of the convulsions which changed one atom into another, it was now clear they involved the conversion of mass–energy into other forms of energy – for instance, into the energy of motion of alpha particles, which shot out of atomic nuclei at tens of thousands of kilometres a second. Ultimately, the extraordinary energy of the atom came from the destruction of mass.

It was a truth which was obvious to Einstein, but not so apparent to everyone else. Most physicists were slow to take on board special relativity, and its disturbing implications for space and time, mass and energy. Francis Aston, however, was different. As he pondered the significance of the anomalously high mass of hydrogen, he came to realise that Einstein and de Marignac had between them provided the key to unlocking the puzzle.

De Marignac had speculated that atoms like helium and carbon would be found to weigh less than the sum of their parts because in their formation mass is lost as energy. Now that Einstein had discovered that energy weighed something the idea made perfect sense. It only remained to explain why energy was lost in the formation of atoms – or, more precisely, why mass-energy was transformed into other forms of energy. Aston saw the answer immediately. The transformation was an inevitable consequence of the force that glued the nuclear building blocks together.

## THE RELEASE OF BINDING ENERGY

The need for a nuclear 'glue' was easy to see. Hydrogen nuclei – protons – repel each other on account of their like electrical charge. However, the fact that heavy nuclei exist implies that there is another force between protons which counteracts the electrical repulsion, preventing nuclei from blowing themselves apart. This force of attraction must become irresistibly strong when hydrogen nuclei get very close together.*

Picture then what happens when hydrogen nuclei approach each other close enough to come under the influence of the powerful nuclear force. Like pieces of shrapnel in an explosion in reverse, they begin to fall towards each other; more and more quickly they fall until, finally, they collide. By the time this happens, however, they have acquired a tremendous energy of motion which they must somehow get rid of if they are to stick together rather than rebound outwards. The surplus energy might be lost in the form of a high-energy particle or gamma

---

* Later to be christened the strong nuclear force to distinguish it from a second, weaker nuclear force.

ray. The details are unimportant. The important thing is that the formation of a nucleus is by necessity accompanied by a loss of energy.

In fact, all objects have less energy bound together than when free. And since energy has mass, they will always weigh less when bound than when free.

The bound system might be the sun plus a meteorite which has crashed into its 'surface'. Energy generated in the impact is lost from the system as heat, so the sun and the meteorite are lighter together than when apart.* The mass loss is of course minuscule, because the gravity binding a meteorite to the sun is weak. Things are very different, however, in the case of an atomic nucleus. The nuclear force binding together its constituents is immensely strong, making the mass deficit significant enough for Aston to have observed with his mass spectrograph.

The obvious question is whether the individual building blocks really are less massive when bound together than when free. Have they somehow shrunk in size? In reality, this is a meaningless question, on a par with the question: 'What time is it?' or 'How long is this ruler?' Just as there is no way to measure time or distance uniquely – it all depends on the motion of the observer – so there is no unique way to measure mass – it all depends on how tightly a body's building blocks are bound together. Space and time, mass and energy: all, by virtue of their interchangeability, are elusive, slippery concepts.

By convention, however, physicists regard the nuclear building blocks as having the same mass inside the nucleus as they do when they are free. The problem then is to explain the loss of energy during the formation of a nucleus. If it doesn't come from the mass-energy of the components, where does it come from?

There is also a second source of energy within the nucleus. Physicists call it 'binding energy'. A nucleus possesses binding energy by virtue of the fact that it is glued together by the nuclear force and it clearly takes energy to pull its components apart. The nuclear force caused the component particles of a nucleus initially to rush together. It gave them their energy of motion, which then had to be shed so that the nucleus would stay glued. The binding energy is therefore the ultimate source of the energy shed by a newly formed nucleus.

When nuclear building blocks come together to make a nucleus, their binding energy must therefore decrease. Low binding energy is therefore synonymous with being tightly bound. However, since the

---

* This was precisely the process proposed by the nineteenth-century scientist Julius Mayer as the source of the sun's warmth.

binding energy of the nucleur building blocks is zero when they are unbound and far apart, the binding energy for a nucleus is always negative. This ensures that the total energy of a nucleus – the mass-energy of its constituents plus its binding energy – is always less than the mass-energy of its constituents. And the more tightly bound the nucleus – that is, the lower or more negative its binding energy – the less it appears to weigh.

The concept of binding energy helped Aston understand the mass deficit he had observed with his mass spectrograph. A nucleus weighed less than expected because in its formation binding energy was inevitably lost.

It all made perfect sense – in theory. But what about in practice? Was there any evidence that the nuclei of a light atom like hydrogen could really come together to form the nucleus of a heavier atom? Remarkably, there was. It was found by Ernest Rutherford the very same year that Aston discovered the anomalously high mass of hydrogen.

## HYDROGEN ATOMS OUT OF THIN AIR

Rutherford owed a great debt to Ernest Marsden, the New Zealander whose alpha-scattering experiment with Hans Geiger had first alerted him to the existence of the atomic nucleus. In 1915, Marsden was continuing his experiments with alpha particles in Manchester. However, instead of firing the tiny bullets at gold atoms, he was now firing them at hydrogen atoms.

Not only was a hydrogen nucleus much lighter than a nucleus of gold, it was also lighter than an alpha particle. When it was struck by an alpha particle, it should therefore recoil at high speed, like a small marble hit by a big marble. Marsden aimed to observe this recoil with a remarkably simple piece of apparatus. It consisted of a small brass box whose air had been pumped out and replaced by hydrogen gas. Sealed inside the evacuated box was a radioactive source which rattled out a stream of alpha particles towards a 'scintillating' screen made of zinc sulphide at one end of the box. In theory, the alpha particles should strike hydrogen atoms in the gas and bounce off them. The net effect of many such collisions would be to transfer energy from the alpha particles to the hydrogen atoms. The alpha particles would slow down and the hydrogen atoms would speed up. Inevitably, some of the speeded-up hydrogen atoms would career off towards the zinc sulphide screen.

When he squinted at the screen through a microscope, Marsden saw the tiny detonations of light that marked the subatomic impacts. All was exactly as he expected. It was when he repeated the experiment, however, that things started to go wrong.

After Marsden had pumped the air out of the brass box, he decided to look through the microscope at the zinc sulphide screen, just to confirm that his apparatus was behaving as expected. Since he had not yet introduced any hydrogen gas into the box, there were no hydrogen atoms to be jolted into motion by the alpha particles. The screen should therefore be totally dark. But when Marsden squinted through his microscope what he saw was not darkness but light. The zinc sulphide was alive with tiny flashes of brilliance. Hydrogen atoms were continuing to hit the screen despite the fact there wasn't the slightest trace of hydrogen in the box. Where in the world could they be coming from?

Marsden could think of only one explanation. The hydrogen atoms must be coming from the radioactive source itself. In addition to firing off alpha particles like miniature firecrackers, the source must be spitting out hydrogen atoms. Evidently, Marsden had stumbled on a new type of particle which was coughed out by radioactive atoms.

The idea was dismissed out of hand by Rutherford. Never, in all his experiments, had he seen the slightest hint of a fourth type of radiation to go alongside alpha, beta and gamma rays. There must be another explanation for the unexpected appearance of the high-speed hydrogen nuclei. And he had a hunch what it was. If he was right, it would make front-page headlines around the world, and everything else he had ever done in science would pale into insignificance. To confirm his hunch, however, he would need to do further experiments.

Had it not been for the Great War raging across Europe, those experiments would certainly have involved Marsden. However, the young New Zealander had exchanged his subatomic bullets for the real thing and gone off to fight on the Western Front – on the opposite side, ironically, from his friend and colleague Hans Geiger. While the Germans tried to starve Britain into surrender by sinking its merchant shipping, Rutherford was working on a programme to develop listening devices for detecting enemy submarines. Consequently, it was only in rare moments that he was able to tackle the mystery of the anomalous hydrogen atoms.

Using equipment similar to Marsden's, Rutherford first proved that the high-speed hydrogen atoms were not coming from the radioactive source. Thus he was able to categorically rule out Marsden's idea that there existed a fourth type of radioactive emanation. Another possibility was that the mysterious hydrogen atoms were coming from the air itself, where hydrogen was known to exist in very small amounts. Not even the world's best vacuum pump could suck out every last whiff of air from a container, so the brass box would undoubtedly retain some residual air.

The air we breathe is a cocktail of gases, which includes nitrogen, oxygen, carbon dioxide, water vapour and even rare gases like helium.* Rutherford began to wonder whether the anomalous hydrogen was associated with any of these. To find out, he began introducing small amounts of each gas in turn into the evacuated brass box.

Oxygen was soon ruled out. When Rutherford leaked it into the brass box, fewer stars twinkled on the zinc sulphide screen than when the box was evacuated. Evidently, hydrogen atoms were slowed down by collisions with oxygen molecules in the box so that fewer made it to the screen. He got the same result when he put carbon dioxide in the box. Next, Rutherford introduced into the box air from which he had squeezed every last trace of moisture. This time, when he bent over the microscope and peered into the eyepiece, he saw that the zinc sulphide screen was twinkling with more stars. In fact, the screen was now being struck by at least twice as many hydrogen atoms as before.

Rutherford had already ruled out oxygen and carbon dioxide as the source of hydrogen atoms. He knew also that the hydrogen atoms could not be coming from water vapour, since he had taken care to dry out the air in the brass box. Of the remaining constituents, the likeliest was nitrogen, a gas which accounted for close to four-fifths of the air we breathe.

Rutherford now introduced pure nitrogen into the brass box. He knew the moment he peered into the microscope that his quest was over. The zinc sulphide screen was alive with even more twinkling stars than it had been when the box had contained air. He had found the source of the anomalous hydrogen atoms. The source was nitrogen. It was a dramatic confirmation of his hunch.

## ASSEMBLING ATOMS

From the day Marsden had mentioned the puzzle, Rutherford had been convinced that something very significant was going on inside the evacuated brass box.

His reasoning was straightforward. The force with which an alpha particle was repelled by an atomic nucleus depended entirely on its

* The helium comes from the decay of radioactive elements in the earth's crust.

electrical charge. For a heavy atom like gold, with 79 protons in its nucleus, that charge was very large, so an alpha particle suffered a violent deflection long before it got to the edge of the nucleus. However, the same alpha particle could penetrate a shade closer to a nucleus with 78 protons, closer still to one with 77, and so on down the list of elements. A sufficiently light element, with only a handful of protons in its nucleus, might offer so little resistance to the express train of an alpha particle that it might smash right through into the nucleus itself.

Rutherford was convinced this was what was happening inside the brass box. The alpha particles from the radioactive source were slamming into the nitrogen atoms in the thin remnant of air. Each time an alpha scored a direct hit, it burrowed deep into the heart of a nitrogen nucleus, triggering violent convulsions. But those convulsions did not spit the alpha particle back out. What came rocketing out instead was a single proton – the nucleus of a hydrogen atom.

The hydrogen atoms in the brass box were more than simply associated with nitrogen atoms. They came from deep inside them. They were chips of the very nucleus itself.

It was a truth which was leapt upon by the world's newspapers when the story broke in 1919. Rutherford had done the impossible, announced their headlines. He had stormed the nuclear fortress itself. He had 'split the atom'.

In fact, the expression downplayed what Rutherford had actually done. For when a proton hurtled out of a nitrogen atom, it left behind a nucleus that was dramatically changed from the one that had existed before. The most common isotope of atmospheric nitrogen, with 7 protons and 7 neutrons, was nitrogen-14. Adding an alpha particle, with 2 protons and 2 neutrons, created a nucleus with 9 protons and 9 neutrons, and subtracting a proton created a nucleus with 8 protons and 9 neutrons. But a nucleus with 8 protons was no longer nitrogen. It was oxygen-17.

Rutherford had therefore achieved what generations of alchemists could only dream of doing. He had taken one kind of atom and artificially changed it into another. What went in was an alpha particle and a nitrogen nucleus, and what came out was an oxygen nucleus and a proton. It turned out that protons were spat out by virtually all elements lighter than calcium when they were bombarded by alpha particles. The only exceptions were helium, carbon and oxygen, whose protons were bound together too tightly to be easily dislodged.

What Rutherford had discovered was an entirely new class of nuclear process. It was the complete opposite of radioactivity. Instead of a heavy atom breaking up into a lighter one, a light atom was built up into a heavier one. Oxygen-17, after all, was heavier than nitrogen-14. Suddenly, there was a precedent for the element-building process envisaged by Prout, in which hydrogen building blocks came together to make heavier atoms – the very same process that Aston's measurements indicated would liberate an enormous amount of nuclear binding energy.

Had Rutherford and Aston only realised it, they now had within their grasp the key to unlocking the secret of the sun and stars. However, neither man saw the wider significance of their two discoveries. It was left to another scientist to put two and two together. That scientist was Jean-Baptiste Perrin, the very same scientist who had used Brownian motion to measure the size of atoms.

## ATOMS AND STARS

It had been clear, ever since Curie and Laborde had astonished the world with their measurement of the extraordinary heat pouring out of radium, that the energy locked inside atoms was perfectly capable of powering the sun for billions of years. The trouble was there was no sign of radium or any radioactive element in the spectrum of the sun.

Now Aston had stumbled on an alternative way that atomic energy might be unleashed – by building up a heavy nucleus from lighter nuclei – and Rutherford had shown that such a process could occur in practice.

Perrin simply put the two pieces together. What Rutherford had achieved artificially in a mundane earthbound laboratory might not nature be able to mimic in a body as extraordinary as the sun? Perrin proposed that the sun constantly transforms hydrogen into heavier elements, and that the nuclear binding energy released in this process was the ultimate source of sunlight. The implications were startling. If Perrin was right, the generation of sunlight was intimately connected with the building up of atoms. Not only did atoms hold the key to understanding the stars but the stars held the key to understanding atoms.

But was Perrin right? It all depended on the conditions in the solar interior. Did the sun, for instance, contain enough hydrogen to fuel Perrin's energy-generating process? The omens did not look good. In 1919 the exact composition of the sun was still a mystery. However, a close examination of the spectrum of sunlight revealed a forest of spectral lines due to the element iron. Far from being made from hydrogen, the sun appeared to be a ball of iron, exactly as Anaxagoras had claimed almost 25 centuries earlier.

Forgetting for a moment the question of whether the sun contained enough hydrogen, there was also the issue of whether the solar interior was hot enough to drive Perrin's atom-building process. In order for hydrogen nuclei to come under the influence of the nuclear force, and 'fuse' together, they would have to approach each other within a million millionth of a millimetre. However, the similar electrical charge of two hydrogen nuclei would cause them to repel each other with a vengeance. The only way they could get close enough together to fuse was if they were slammed into each other at an extraordinarily high speed.

Ultra-high speed meant ultra-high temperature. The sun's interior would have to be extremely hot. In fact, it was possible to calculate how hot a gas of hydrogen nuclei would have to be in order that the nuclei could get close enough to fuse together. The temperature was an unbelievable 10 billion degrees. Was it conceivable that such extraordinary temperatures existed in the interior of the sun?

Since there was no chance of travelling to the sun, let alone sticking a thermometer into its depths, how could we ever learn the conditions inside it? Incredibly, an English astronomer found a way.

# 6

# The Infernal Constitution
# of the Stars

HOW WE DISCOVERED THE EXTRAORDINARY
CONDITIONS INSIDE THE SUN AND STARS,
BUT REALISED THAT THE TEMPERATURE
WAS FAR TOO LOW TO CONVERT
HYDROGEN INTO HELIUM.

Twinkle, twinkle, little star,
how I wonder what you are.

Jane Taylor

A star that burns twice as brightly burns half as long.

Roy the android (*Blade Runner*)

Arthur Stanley Eddington was a slender, nervous-looking man who gave the impression of being totally ill-equipped for the rough and tumble of combative science. He was cripplingly shy. And, on the rare occasions when he plucked up the courage to speak, what emerged was completely incoherent. As one of his Cambridge students put it, 'His brain had no proper connection with his mouth.' But although Eddington was renowned as the world's worst lecturer, a man who never finished a sentence before starting another, paradoxically he wrote with extreme clarity for both academic and popular audiences. It was an ability that enabled him to win most of his scientific battles, and which also led to his international fame.

That fame was the result of a report he was asked to write on the general theory of relativity, Einstein's extension of special relativity to include the effects of gravity. The report, published in 1918, gave the English-speaking public its first opportunity to learn the details of the new theory, which had been completed three years earlier at the height of the First World War. More importantly, it turned Eddington into an expert on general relativity and its predictions, principally the light-bending ability of gravity.

Einstein's theory predicted that if the light from a distant star passed close to the sun on its way to earth, its trajectory should be bent about twice as sharply as Newton would have predicted. Such an effect would cause the position of a star to shift slightly relative to other stars. Though impossible to see in the glare of daylight, it might be observable during a total eclipse when the moon blotted out the bright solar disc. Such an eclipse was due to occur on 29 May 1919 and Eddington travelled to the island of Principe, off the coast of West Africa, to see it. When his photographs confirmed that starlight was indeed deflected by the amount predicted by general relativity, he earned his place in the history books as 'the man who proved Einstein right'.

But despite Eddington's contribution to general relativity, his principal scientific interest lay elsewhere. Ever since he had started out in astronomy as a 24-year-old in 1906, he had had one overriding goal: to understand the inner workings of stars.

## BALANCING ACT

The goal, at first sight, appeared an impossible one. There was no hope after all of ever looking inside a star. Nevertheless, it was possible to say some very general things about stars like the sun. And this is what Eddington proceeded to do.

The sun is held together by gravity, which is trying to pull every piece of the sun down towards its centre. If gravity ever got its way, it would crush the sun to a mere speck in less than an hour, turning it into a black hole.* The fact that such a catastrophe has never occurred is because there is another force in the solar interior which is opposing gravity. That force is generated by the stuff of the sun itself, which, like the material of a rubber ball, resists being squeezed.

The resisting force does more than prevent the sun from shrinking to form a black hole, it prevents it from shrinking in any way at all. From this apparently trivial observation, Eddington deduced that at every

---

* The gravity of a black hole is so strong that nothing, not even light, can escape, hence its blackness.

point in the solar interior the inward force of gravity is perfectly balanced by the outward force of the solar material pushing back.

One way to think of this balance is to imagine a thin spherical shell of matter inside the sun. On the inside of the shell is a pressure pushing outwards; on the outside surface is a pressure pushing inwards. Of these, the pressure pushing outwards must be the greater since it must not only balance the pressure pushing inwards but also the force of gravity trying to drag the shell inwards.[*]

The sun can be thought of as made of many such concentric shells, rather like the multiple skins of an onion. Since the pressure on the inside of each shell must be greater than the pressure on the outside, the pressure of the material of the sun must rise steadily towards the centre, reaching a maximum at the centre itself.

The pressure on the outside of any shell can loosely be thought of as the weight of the material bearing down on it from above. The pressure at the centre of the sun is therefore the same as the weight of the entire sun. This is a truly tremendous pressure, equivalent to 100 million tonnes resting on every square centimetre.[†]

Clearly, an understanding of the inner workings of the sun is impossible without a knowledge of how matter behaves under such extraordinary pressures. On average, the sun is denser than water, which means that in the core, where the conditions are most extreme, the material must be denser than a typical solid. Solids resist being crushed because they are inherently 'stiff'; try crushing a rock in your hand and feel how hard it pushes back. The stiffness of rocks enables those at the centre of the earth to support the entire weight of the planet bearing down from above. But the weight pressing down on the centre of the earth is a mere ten thousandth of the weight pressing down on the centre of the sun. There is no way that any normal solid can possibly be stiff enough to withstand such a weight.

Most people, confronted by this truth, would have thrown their arms up in despair and given up trying to understand the innards of stars. Not so Eddington. It was his great genius to realise that there was a class of stars whose interiors were nowhere near as extreme as the sun's and which, conceivably, he might be able to understand.

## GOODNESS GRACIOUS, GREAT BALLS OF GAS

The stars he had in mind looked for all the world like dying embers.

---

[*] The force of gravity on any body is simply another word for its weight.

[†] Or, to put it another way, 20 million elephants resting on a penny!

They had been recognised as a distinct class by the American astronomer Henry Norris Russell, who had used their colour to deduce their temperature. Just as a red-hot fire is cooler than a yellow-hot one, a red star is cooler than a yellow one. Russell's stars were therefore cooler than the sun, with surfaces closer to 3000°C than the solar temperature of 6000°C.

There is nothing unusual about this. Many stars are cooler than the sun. However, none are quite as extraordinary as Russell's. Common sense said that a cool star should also be a dim star. However, Russell's stars were pumping out thousands, or even tens of thousands, of times as much heat and light as the sun. Far from being dim, they were fantastically luminous.

There was only one way a star could be both cool and prodigiously luminous: by being big. A household radiator gives out more heat than a white-hot sparkler because it has a far greater area. In the same way, Russell's peculiar red stars could pump out more heat than hotter stars like the sun only if they had very large surface areas.

Russell christened the stars 'red giants'. They were so puffed up and bloated that if a typical member of the class were to take the place of the sun it would fill most of the space inside the earth's orbit. Viewed from the earth's surface (or the blackened slag that would pass for the surface) it would be a ghastly swollen sun, covering more than half the sky and drowning the landscape in blood-red light.

Some red giants are so monstrous, in fact, that they would not even fit inside the earth's orbit. If the sun were to be replaced by Betelgeuse in the constellation of Orion, for example, or Antares in Scorpio, the earth would actually find itself inside a star. Since the amount of matter in a red giant is typically the same as in the sun – it has simply been smeared throughout a tremendously bigger volume – the earth would be ploughing through angry red star stuff little more substantial than thin air.

It was the ethereal nature of red giants which greatly impressed Eddington. A red giant was nothing more than a giant ball of gas. And this, he realised, was the key to understanding such stars; for a gas was a gas was a gas, irrespective of whether it was confined beneath a piston on earth or within the globe of a vast red giant.

Somehow the gas in the core of a red giant must generate the tremendous pressure required to support the weight of the star. The pressure of perfect gas – the kind of gas first imagined by Daniel Bernoulli – goes up if it becomes more dense, in which case the gas particles hammer more frequently on the walls of a container. It also goes up if the gas becomes hotter, in which case the gas particles also hammer harder and more frequently. If, therefore, the gas in the core of a red giant is to generate an extremely high pressure, it must be both dense and hot.*

But how hot? It was obvious to Eddington that the temperature would have to be extraordinarily high by everyday standards. To obtain a precise figure, however, he needed to know the exact pressure and exact density at the centre of a red giant. This, in turn, would require a more sophisticated picture of the interior of such a star.

## GOODNESS GRACIOUS, GREAT BALLS OF LIGHT

A red giant is considerably more than a ball of gas in which the inward force of gravity is balanced by the outward force of the hot gas. It is a ball of gas which is also leaking heat into space.

Usually, when a body loses heat, it gets colder. In the case of a gas, this causes a drop in pressure. If the gas happens to be in a red giant, however, the drop in pressure will upset the delicate balance of forces deep inside. Gravity will get the upper hand and the star will start to shrink.† The fact that a red giant continually loses heat to space without cooling or shrinking shows that some other process is at work to maintain the star's internal temperature, so that gravity never gets the upper hand. The process is the flow of heat from the deep interior of the star up to the surface, a flow which exactly replenishes the heat lost to space.

Such a heat flow occurs for the simple reason that the centre of a red giant is hotter than its surface, and heat always flows from a hot place to a colder place. The same thing happens when one end of a metal spoon is dipped into a boiling saucepan and heat flows towards the other end. However, there is a crucial difference between heat flowing along a metal spoon and heat flowing outward through a gaseous red giant, a difference which Eddington was the first person to appreciate.

To understand the difference, it is necessary to know that heat energy in a metal exists in two basic forms. These are particle motion – the incessant jiggling of atoms, and radiant energy – better known as light.‡

---

* Strictly speaking, of course, it's the difference in pressure across a shell of material and not the pressure that counterbalances the force of gravity.

† Bizarrely, if a ball of gas shrinks in this way, gravitational energy is converted into heat through internal friction in the gas, and the gas actually heats up. In other words, when a gas ball loses heat it gets hotter. This is in marked contrast to a lump of coal which gets colder when it loses heat.

‡ Strictly speaking, this is infrared light: a kind of invisible light that warms your skin when you face the sun.

Of the two forms, radiant energy is the least important. It accounts for a tiny fraction of the total heat content of a metal. However, as a body gets hotter, the energy in radiation grows at a faster rate than the energy in particle motion. Radiant energy therefore becomes ever more significant at high temperatures.

For a metal, the consequences are minor – even when a metal bar is red-hot, the radiant energy inside still accounts for a mere billionth of its total heat energy. For a red giant star, however, the consequences are very important indeed, because of the star's insubstantial nature. For, even though the gas particles are flying about very quickly, they are so few and far between that the total energy tied up in their motion is much less than the energy of the atoms jiggling in a similar volume of metal. With the heat energy of matter so diluted in this way, radiation takes on a far greater importance. Eddington was the first to realise that radiation accounts for a significant proportion of the heat energy of one of Russell's stars. From the point of view of heat, a red giant is nothing more than a giant globe of bottled-up light.

This key insight has profound implications for our understanding of the way in which heat flows out from the hot centre of a red giant. In a metal the flow occurs when one metal atom jiggles the next, rather like a fidgeting child infecting her classmates. However, in a red giant, full of bottled-up radiation, heat flows up from the hot core to the cool surface in the form of light.

It is a painfully slow process. Although the light from this page flies directly to your eye, the light escaping from the hot interior of a red giant is constantly obstructed by gas particles which block its way like an unhelpful crowd of pedestrians. Consequently the light ricochets back and forth, travelling only a short distance before being sent off in another direction. Indeed, the zigzag path followed by a typical light ray inside a red giant is so tortuous that it may take hundreds or even thousands of years to work its way to the surface and attain the freedom of outer space.* So, although red giants give the impression of being spendthrifts with their light, in reality they are misers, holding on to their great store of internal heat and leaking it into space at the barest of trickles.

The trickle of light that escapes is very different to the light that started its journey deep down in the star. At the centre, where the temperature is at its highest, it is in the form of high-energy X-rays. But

---

* Precisely how long depends on the 'opacity' of the material, which is merely a technical word for how good it is at impeding the flow of light. In the sun, light takes about 30,000 years to work its way from the centre to the surface, a journey that would take just over 2 seconds if its path was a straight line.

as it threads the endless maze of the star, collisions with gas atoms continually sap the light of energy. This process helps maintain the temperature of the gas in the face of the constant leakage of heat into space. But it means that by the time the light emerges from the surface of the star, it has been degraded to much lower-energy red light.

Eddington's intellect had laid bare the interior pressures of a red giant. But there was a long way to go. If he were to succeed in his goal of really understanding such stars, he would need to work out the precise consequences of those processes. For a red giant of a given mass, surface temperature and size, he therefore asked himself how the pressure, density and temperature would have to fall off with distance from the middle of the star in order to ensure that gravity was perfectly balanced by the outward force of the hot gas at every point inside.*

However, complications arose, because the force of the hot gas depended on its temperature, which was maintained by the flow of heat outwards from the centre. This in turn depended on gravity, since the greater the weight of the star pressing down on the core, the greater the temperature, and the faster heat flowed out of its core. Everything, in other words, depended on everything else. It was enough to make anyone's head spin. Anyone's head, that was, except Eddington's.

Eddington presented his theory of the 'radiative equilibrium of stars' to the Royal Astronomical Society at the height of the First World War in 1916. He had by now worked out the central temperature of a red giant: it was a staggering 7 million degrees. The reception his paper received was encouraging, though not over-enthusiastic. Red giants were, after all, a relatively minor class of star. And his theory had nothing to say about the huge majority of ordinary stars, which of course included the sun.

How wrong everyone was. Eddington's theory would turn out to have wider significance than he, or anyone else, had suspected. Paradoxically, this realisation came about as a result of a criticism levelled at the theory. After the presentation at the Royal Astronomical Society, several scientists pointed out that in concocting his theory Eddington had failed to take into account the most up-to-date picture of atoms.

---

* In fact, Eddington believed that the pressure to balance gravity came not only from the hot gas but also from the light flowing outwards through the star. Light exerts a pressure because it carries energy, and energy, according to Einstein, has an equivalent mass. Nowadays, however, we know that such 'radiation pressure' is important only in very massive stars which are very hot and luminous.

## THE WEIRD WORLD OF THE ATOM

The new picture of atoms had arisen to overcome a fundamental flaw in the picture conceived by Ernest Rutherford in 1911. It concerned the electrons which flitted about an atomic nucleus like planets round the sun. According to the theory of electromagnetism, a charged particle gives out light when it changes either its speed or direction of motion. It follows that an electron circling a nucleus should act like a miniature transmitting station, constantly broadcasting light waves into space. However, the light it radiates will sap an electron of its energy of motion, causing it to spiral into the nucleus. In fact, calculations showed that it would collide with the nucleus within a mere hundred millionth of a second.

By rights, atoms should not exist.

The man who came up with a better picture of the atom was the Danish physicist Niels Bohr, who arrived in Manchester to work with Rutherford in 1912. Bohr reasoned that since the known laws of physics forbade the existence of atoms – yet atoms clearly did exist – the known laws of physics simply did not apply in the atomic realm. Atoms danced to the beat of an entirely different drum. It was one of the most breathtakingly audacious suggestions in the history of science. Freed from the shackles that constrained every other physicist, Bohr began to devise a new picture of the atom which was in accordance with the known behaviour of atoms. The picture he came up with in 1913 is best visualised for the simplest atom, hydrogen, in which a single electron orbits a nucleus consisting of a solitary proton.

Bohr maintained that there are special orbits in an atom. If an electron occupies such an orbit, it can circle endlessly without radiating light in all directions. The orbits exist at a limited number of precise distances from the nucleus. This is like saying that a planet might orbit the sun at, say, 30 million miles or 60 million miles away but not at 20 or 45. Bohr further postulated that there existed an innermost orbit, inside which an electron could not venture. By these apparently arbitrary means, he was able to stabilise his atom against the catastrophic shrinkage that plagued Rutherford's atom. However, Bohr had not the slightest idea of the atomic laws which forced an atom to be the way he imagined it. (He hoped to discover them at some later date.) His only justification for the new picture of the atom was that it explained things.

Since the electron in a hydrogen atom was gripped by the electric force of the nucleus, it would have to be supplied with energy to jump from an orbit close to the nucleus to one further out. To jump down from a faraway orbit to one close to the nucleus it would have to shed energy. It was Bohr's suggestion that when this excess energy was shed it was shed in the form of light.

Immediately the new model atom made sense of the light emitted by a hydrogen atom. Since there were a limited number of possible orbits available to the electron, there were a limited number of possible jumps from outer orbits to inner orbits. Each jump resulted in the emission of light of a distinctive energy – the energy difference between the two orbits. Since the energy of light was related to its wavelength – with shorter wavelengths having higher energy than longer wavelengths – each jump between possible orbits resulted in the emission of a characteristic wavelength of light. After a century of research, someone had at last come up with an explanation for spectral lines.

An immediate success of Bohr's scheme was that it made sense of a curious pattern in the wavelengths of the spectral lines of hydrogen. The pattern had been noticed in 1888 by Jacob Balmer, an ageing teacher at a Swiss girls' school. Like the pattern in Mendeleev's Periodic Table, it strongly hinted that atoms were made of smaller things. The 'Balmer series' of spectral lines could be explained, Bohr realised, if the light corresponding to each line was shed when an electron in a hydrogen atom dropped down to the second innermost orbit. Electrons might drop from the third orbit to the second, from the fourth to the second, from the fifth to the second, and so on. In each case the hydrogen atom would emit light with a wavelength which corresponded to the energy difference between the two orbits.[*]

Hydrogen is the simplest atom, and things of course become more complicated with heavier atoms with more than one electron. Each electron has its very own orbit, which it does not share with any other, and is free to jump down to an orbit nearer the nucleus only if that orbit is not already occupied by another electron.[†] Since in an atom with, say, 10 electrons, the electrons tend to fill up the ten innermost orbits, this restriction is a severe one.

But although electrons tend to circle nuclei obediently in their prescribed orbits, there are circumstances in which one or more of them could be kicked out of an atom altogether. It was this process, known as 'ionisation', which turned out to have profound implications for Eddington's theory of red giants.

---

[*] Hydrogen also emits a characteristic series of 'Lyman' lines when its electron drops down to the innermost orbit. However, these spectral lines are in the ultraviolent region of the spectrum and so invisible to the naked eye.

[†] In fact, this is not quite true. It would become clear later that two electrons could share a given orbit.

## THE SUN IS A GAS

Since every electron in a particular atom occupies a precise orbit, it follows that a precise amount of energy is needed to tear it away from the nucleus – the 'ionisation energy'. If an electron receives less than the ionisation energy, it will stay trapped around the atom; if it receives more, it will escape to freedom. This ionisation energy is smallest for an atom's outermost electrons, since they circle far from the pull of the nucleus, and largest for those that orbit close in, where the pull is strongest.

The most important feature of this was that each electron had its own unique ionisation energy. Consequently, if all the electrons in an atom were to be somehow supplied with, for example, 10 units of energy, those with ionisation energies below 10 would be ejected while those with ionisation energies above 10 would be unmoved.[*]

In fact, there is indeed a situation in which every electron receives, roughly speaking, the same quantity of energy: in a dense, hot gas. The hotter the gas, the greater the average energy of motion of its atoms (or, equivalently, the faster they are moving). One way this energy of motion can be transmitted to atomic electrons is during violent collisions between atoms.

If the temperature of a gas is known, it is possible to calculate the average energy of motion of its atoms. If this energy is greater than the ionisation energy of, say, the outermost electron of each gas atom, during collisions most gas atoms will lose their outermost electrons. Similarly, if this energy is greater than the ionisation energy of the second outermost electron, most gas atoms will have these electrons ejected as well, and so on.

And, here we come to the criticism which was levelled at Eddington's theory of red giants by his astronomer colleagues. He had assumed that the gas from which such stars were made consisted entirely of neutral atoms – atoms with their full complement of electrons. However, Eddington's critics pointed out that the gas in a red giant was at an extremely high temperature. The gas atoms were therefore very unlikely to be neutral. In fact, it was probable that most had already lost many of their electrons. Far from being a normal gas of atoms, as Eddington had assumed, the matter deep inside a red giant would be a stew or 'plasma' of electrons torn from atoms, plus the charged nuclei, or 'ions', created in the process.

Despite consisting of ions and electrons, a plasma is still a gas. It

---

[*] In practice, electron energies are measured in peculiar units known as 'electronvolts'.

therefore obeys the law of a perfect gas. Where it differs from a normal gas – apart from in the obvious way of being electrically charged – is in the number of particles flying about. A box containing a plasma of a million iron ions and their lost electrons will clearly contain more particles than a box containing a million iron atoms.

Eddington had assumed that red giants, like the earth, were made mostly of iron. He had furthermore assumed that the iron was in the form of neutral atoms. However, in the light of the latest thinking on atoms, this could not be true. For, at a central temperature of 7 million degrees, each iron atom would have lost about half its total complement of 26 electrons.

Having more particles flying about affects the pressure of a gas, since the pressure at a given temperature and density depends on how many particles there are hammering on the walls of a container: double the number and the pressure is doubled. A box of a million iron ions and their electrons therefore exerts more pressure than a box containing a million iron atoms at the same temperature. Looked at another way, an iron plasma can generate the same pressure at a lower temperature and density than a gas of iron atoms.

The significance for a red giant was clear. Since its core was an iron plasma rather than a gas of iron atoms, it could support the overlying layers of the star with a lower density and temperature. Eddington carried out the necessary calculations and concluded that the density need only be a tenth the density of water. His revised estimate of the central temperature of a red giant was 5 million degrees.

By incorporating the new ideas about atoms into his theory, Eddington had silenced his critics. However, he was still no nearer to his ultimate goal – understanding ordinary stars like the sun. During the 1920s, he began to wonder whether one way ahead might be to simply modify his theory of red giants.

## THE INFERNAL CONSTITUTION OF THE SUN

Eddington began by trying a different gas law. This was the law of an 'imperfect' gas, which had been formulated by the Dutch physicist Johannes van der Waals. The law described the behaviour of a gas which was squeezed so hard that its atoms were literally jammed into each other. It was highly unlikely that solar material obeyed van der Waals' law. Nevertheless, Eddington out of desperation assumed that it did. When he calculated the density of the centre of the sun using this assumption, he came up with a figure of 13 times that of water. The corresponding temperature was 18 million degrees. It was the first real indication of the conditions in the very heart of the sun.

At around the same time, Eddington persuaded the mathematical physicist Arthur Milne, a Cambridge colleague, to take a fresh look at the ionisation of gas in stars. By now, Bohr's rather arbitrary model of the atom had been put on a firmer theoretical footing and it was possible to determine with more confidence the state of atoms deep in the core of a red giant. Milne confirmed that at a density of a tenth that of water and a temperature of 5 million degrees, *all* atoms, not just those of iron, would be stripped of most of their electrons. However, buried in Milne's work Eddington found something far more important than a simple verification of what he already knew. His calculations appeared to predict that even if a gas was squeezed to a higher density than in a red giant, the number of electrons torn from a typical gas atom would hardly change. In fact, matter would still be in the form of a plasma, with the atoms clinging on to a few tightly bound electrons, even at a density hundreds of times higher than that found in the core of a red giant.

Eddington was stunned. The density at the centre of the sun was hundreds of times that in the heart of a red giant. Remarkably, even though the matter there was denser than a solid, it was still in the form of an ionised gas. His theory of the interior of red giants therefore applied equally well to the interiors of ordinary stars like the sun. He had been seeking a theory that would explain all stars, not just a minor subset of them. Suddenly, and unexpectedly, it had fallen into his lap.

The ultimate test of any theory is of course how well its predictions match observations. Eddington's theory made a very specific prediction about the light output, or 'luminosity', of a star. The luminosity – at least for stars with the same chemical composition – should depend solely on their mass. This dependence was very sensitive; if a star was, for example, twice as massive as another, it was not simply twice as luminous but about twenty times as luminous.

Eddington looked up all the stars whose masses and luminosities had been measured. Using the mass of each star, he predicted its luminosity. The result was stunning. In every case, the predicted luminosity agreed with what astronomers had measured.

In 1924, he presented his results to the Royal Astronomical Society. Shortly afterwards, he began expounding his theory in book form. *The Internal Constitution of the Stars* became an instant classic when it was published in 1926. Eddington's theory of stars was a remarkable achievement. Despite the impossibility of tunnelling inside stars to see what their interiors were like, he had managed to lay bare their inner workings. By applying his theory to the sun, he concluded that the matter at the centre is a staggering 77 times as dense as water, and that its temperature is 40 million degrees.

The most remarkable feature of Eddington's theory of stars, however, was that it had been formulated without any reference to the ultimate source of stellar energy. It was Eddington's genius to realise that a star is hot not because of any power source supplying it with heat, but because of the mass bearing down on its core.

This is only possible if the stellar energy source generates heat at whatever rate is needed to maintain the *status quo* of a perfectly balanced ball of hot gas. This means that the energy source must have its own built-in thermostat so that if it generates too little heat and the core gets too cold, it responds almost instantly by generating more heat. Similarly, if it generates too much heat and the core gets too hot, it must respond just as quickly by reducing the energy it is generating.

When the core gets too cold, it shrinks and heats up; when it gets too hot, it expands and cools. The energy source must therefore step up its heat output in response to a small rise in temperature and decrease it in response to a small drop in temperature. In short, it must be extremely responsive to the temperature.

All this Eddington was able to deduce from the mere fact that many of the properties of stars did not depend on what was powering them. But though his theory did not require him to know what was replenishing the heat flooding from stellar cores, he nevertheless spent a great deal of time wondering about what was powering the sun and stars.

## A HYDROGEN-POWERED SUN?

Ever since 1903, when Pierre Curie and Albert Laborde stunned the world with their discovery of the fantastic heat output of radium, Eddington had been convinced that the stars were powered by 'subatomic' energy. However, the possibility that this energy was released by radioactivity had been ruled out when astronomers failed to find the spectral signature of radium – or uranium or thorium – in the solar spectrum. Eddington was forced to conclude that subatomic energy in the sun was being unleashed by some, as yet unknown, process.

Whatever the energy-generating process, it was likely to require extreme conditions of a kind not found on earth. Since the most extreme conditions in the sun existed deep in its heart, this was where

Eddington believed the sun was generating its energy, not on the relatively cool surface where astronomers had looked for signs of radioactive elements. One possibility, vigorously promoted by Eddington's Cambridge colleague James Jeans, was that subatomic energy was released in the sun during the 'total annihilation' of matter. This was a hypothetical process in which 100 per cent of the mass-energy of matter was transformed into other forms, ultimately heat-energy.

In order to maintain the sun's tremendous energy output it would be necessary to annihilate 4 million tonnes of matter each second. Although on the face of it this appeared to be a tremendous amount of matter, in fact it was only a tiny, tiny fraction of the 2 thousand million million million million tonnes contained within the sun. If total annihilation were the ultimate source of solar energy, there was enough matter to keep the sun blazing not merely for billions of years but for thousands of billions of years.

The Achilles heel of Jeans' idea was that there was no evidence that the building blocks of matter – protons and electrons – could really vanish in a puff of radiant energy.* This lack of evidence was not enough to cause Eddington to rule out completely total annihilation as the energy-generating mechanism of stars. However, his instinct was to plump for a process supported by at least a modicum of experimental evidence. He was therefore overjoyed when he first heard Jean-Baptiste Perrin's proposal that stellar energy might be a by-product of the conversion of hydrogen into heavier elements. Here, surely, was the power source of the stars.

Not only had Francis Aston shown that mass-energy would be converted into other forms during the process of atom-building but Ernest Rutherford had actually succeeded in transforming a light element into a heavier one in the laboratory. 'And what is possible in the Cavendish Laboratory,' Eddington declared in 1920, 'may not be too difficult in the sun.'

The simplest conceivable atom-building process, and the one favoured by Eddington, was the one in which hydrogen atoms were converted into atoms of the next heaviest element, helium. Aston's measurements had indicated that, during such a transformation, 0.8 per cent of the mass of every hydrogen atom would vanish and reappear as another form of energy. Since the process was less than a hundredth as

---

* In fact, total annihilation is possible. However, it occurs only when particles such as protons and electrons meet their respective 'antiparticles' (the first antiparticle, the 'positron', was discovered in 1932). Unfortunately, it is impossible to imagine a plausible scheme in which total annihilation generates a significant amount of energy because of the difficulty of generating large numbers of antiparticles.

efficient as total annihilation, the maintenance of sunlight required the consumption not of 4 million tonnes of hydrogen a second but more than 400 million tonnes a second.

The sheer scale of hydrogen–burning in the sun dwarfed the imagination. However, according to Eddington's calculations, the sun could easily burn for 10 billion years if hydrogen amounted to as little as 7 per cent of its total mass – as long again as the sun had already been shining. This percentage of solar hydrogen was higher than Eddington would have liked, wedded as he was to the idea of an iron sun, but he could live with it.

At last the puzzle of what was making the sun shine appeared to be solved. All the pieces seemed to hold together. All the pieces, that is, except one. In order for two hydrogen nuclei to get close enough together to 'fuse', they would have to be slammed together at ultra–high speed. Ultra–high speed means ultra–high temperature. In fact, the temperature would have to be about 10 billion degrees. But, according to Eddington's theory of stars, the temperature at the heart of the sun was only 40 million degrees. Infuriatingly, the centre of the sun was nowhere near hot enough to weld hydrogen into helium.

Despite this setback, Eddington remained convinced that the conversion of hydrogen to helium was the only conceivable energy source of the sun. When other astronomers chided him for ignoring his own evidence that the sun was too cold, he responded: 'We do not argue with the critic who urges that the stars are not hot enough for this process; we tell him to go and find a hotter place.' In other words, go to hell!

There was only one way out of the dilemma. Somehow the fusion of hydrogen into helium in the sun must be going on at a far lower temperature than 10 billion degrees. But how in the world was that possible? A flamboyant Russian physicist called George Gamow provided the answer.

# 7

# *Tunnelling in the Sun*

## HOW WE FOUND THAT THE FUSION OF HYDROGEN IN THE SUN WAS POSSIBLE AFTER ALL, AND IDENTIFIED HOW THIS MIGHT HAPPEN.

That evening, after we had finished our essay, I went for a walk with a pretty girl. As soon as it grew dark the stars came out, one after another, in all their splendour. 'Don't they shine beautifully?' cried my companion. I simply stuck out my chest and said proudly, 'I've known since yesterday why it is that they shine!'

Fritz Houtermans

George Gamow was a flamboyant character with a reputation for drinking, womanising and generally wringing the maximum from every day of his life. He was also one of the most intuitive and inventive scientists of the twentieth century.

Gamow was born in the Ukrainian port of Odessa in 1904. He became interested in science when as a boy he began to wonder about the truth of the Christian dogmas taught in school. Was it really possible that the wine and bread served during Communion could change into the blood and flesh of Christ? There was only one way to find out. Gamow examined samples of transubstantiated wine and bread under his small microscope. When he saw no sign of any change, he was profoundly affected. 'I think this was the experiment that made me a scientist,' he later claimed.

In Leningrad, Gamow studied physics under Aleksandr Friedmann, the man who first guessed that the universe was expanding in the aftermath of a 'big bang' explosion. In 1928, on gaining his doctorate, the 24-year-old Gamow obtained permission from the Soviet authorities to attend a summer school at the University of Göttingen. It was during his two-month stay in the German town that he made the crucial discovery which hinted at how hydrogen might be converted into helium even at the relatively low temperature in the sun.

The problem which stimulated Gamow to make his discovery appeared at first sight to have nothing whatsoever to do with generating energy inside stars. Instead it concerned the emission of alpha particles by radioactive atoms.

## THE PECULIAR PUZZLE OF ALPHA DECAY

By the 1920s, the process of alpha decay had thrown up a baffling puzzle. Alpha particles appeared to be able to smash their way out of the nuclear fortress even when they had insufficient energy to do so.

The best way to understand this puzzle is to think of the force field barrier surrounding a typical atomic nucleus. Rutherford's alpha-scattering experiments – which had identified the nucleus in the first place – had revealed that the closer an alpha particle got to a nucleus, the more fiercely it was repelled. This repulsion was not difficult to understand. It is simply due to the electrical force which drives apart any two particles which had a similar electrical charge – in this case the positively charged alpha particle and the positively charged nucleus.

But Rutherford's experiments had also revealed something else. When an alpha particle passed very close indeed to a nucleus, the electrical repulsion it experienced was slightly less than expected. It was the first tantalising evidence of an entirely new force which cancelled out the electrical force close to the nucleus.

The existence of such a force had long been suspected by physicists. After all, the tremendous electrical repulsion between protons should in theory blow a nucleus apart. Without some kind of 'nuclear force' to counteract this repulsion and glue protons and neutrons together, there was no way a nucleus could exist.

The essential features of the force field around a nucleus can be neatly captured by a simple model: a toy volcano on a table-top. The slopes of the volcano represent the electrical repulsion around the nucleus, and the crater at the summit corresponds to the nuclear force of attraction which glues together the particles of the nucleus.

Now imagine rolling a marble across the table-top towards the volcano. This is the same as firing an alpha particle at a nucleus. If the marble lacks the necessary energy to reach the lip of the crater, it rolls back down to the table-top. In much the same way, an alpha particle with insufficient energy to overcome the repulsive barrier of a nucleus ricochets off that nucleus. This may all seem very obvious. Where the table-top model comes into its own, however, is in illuminating the central puzzle of alpha decay.

Before being spat out, an alpha particle is imprisoned inside a nucleus, like a marble rolling to and fro in the crater of the volcano. In order to break free, the marble must surmount the lip of the crater. If it can do this, it will roll away down the slope, speeding up all the while. By the time it reaches the table-top, it will have gained a considerable energy of motion.*

This gain of energy is unavoidable. For an alpha particle escaping from a nucleus of uranium-238 it amounts to 17 megaelectronvolts (MeV).† No alpha particle emerging from uranium-238 can have less energy than this. Imagine, then, the consternation of Rutherford and others when they measured the energy of alphas coming from uranium-238 and found that it amounted to just 4.2 MeV – less than a quarter of the energy it ought to be.

Rather than rocketing out of a uranium nucleus, an alpha particle was merely dribbling out. It was as if a marble, rattling about in the crater of the toy volcano, had magically appeared three-quarters of the way down the slopes and quietly rolled the remaining distance down to the table-top. It made no sense. No sense, that was, until Gamow gave the puzzle some thought.

The mistake everyone was making, Gamow realised, was in expecting alpha decay to conform to common sense. Although common sense undoubtedly applied in the everyday world, as Niels Bohr had realised, it emphatically did not apply in the realm of the very small.

## WAVING GOODBYE TO CERTAINTY

One of the first indications that the world of atoms was profoundly different from the everyday world of our senses had come from

---

* In fact, it is precisely the energy of motion it would have required to roll up to the crater rim from the table-top.

† Just as the energies of orbiting electrons are measured in terms of the peculiar unit known as the 'electronvolt' (eV), the energies of nuclei are measured in terms of millions of electronvolts or 'megaelectronvolts' (MeV). This reflects the fact that, when the constituents of nuclei are rearranged, the quantities of energy involved are a million times greater than those associated with rearrangements of electrons. Thus a nuclear energy, which involves the rearrangement of a nucleus, is a million times as powerful as a chemical energy, which merely involves the rearrangement of orbiting electrons.

radioactivity itself. All the atoms of a radioactive element such as radium are identical, just as all the atoms of a nonradioactive element like gold are identical. It is this identicalness, after all, that defines an element, that makes it elemental. However, some atoms in a sample of radium disintegrate more quickly than others.

Here was a conundrum. How could radium atoms all be the same yet behave differently?

When a radioactive substance decayed, it did so according to a simple law discovered by Rutherford: the time taken for half the atoms in a sample of radioactive material to decay is always the same. This time interval is unique for each radioactive substance, and was christened its 'half-life'. If a material's half-life is 10 minutes, for instance, half the atoms in a sample will remain undecayed after 10 minutes, a quarter after 20 minutes, an eighth after 30 minutes, and so on.

Superficially, this characteristic may seem perfectly innocuous. However, it is actually extremely peculiar. The only way it can arise is if in any given interval of time each atom in a radioactive sample has precisely the same chance, or 'probability', of self-destructing. In other words, radioactive decay is governed by pure chance like the throw of a dice or the spin of a roulette wheel.

This statement takes a while to really sink in. It means that nature forbids us, even in principle, from predicting when an individual atom in a radioactive sample will decay. All we can know is the probability that it will decay in some time interval.

The conundrum of radioactive decay is neatly solved. For if only the probability of an individual atom in a sample decaying is knowable, and that probability is the same for all atoms, then every atom, whether it chooses to decay after 10 seconds, 10 minutes or 10 hours, is to all intents and purposes the same. In the microworld, it turns out, 'identical' has a very different meaning.

Having to accept probabilities in the place of the certainties was a bitter pill for physicists to swallow. However, this was merely the first of many bitter pills. For, in time, it became clear that the process of radioactive decay was in no way unique. Virtually every aspect of atomic behaviour was ruled by pure chance.

It is impossible, for instance, to know that an electron has a particular velocity, or occupies a particular orbit in an atom. All that can be said is that there is a particular probability that it might be found with that velocity, or a particular probability that it might lie in that orbit. These probabilities could of course be zero or 100 per cent, but in general were somewhere in between. Everything about atoms turned out to be vague and uncertain.

However, nature, having robbed us of certainty, handed us a consolation prize. Although we were forbidden to know anything about atoms with 100 per cent certainty, we were nevertheless permitted to know the probabilities with certainty.

Bizarrely, the behaviour of particles of matter was found to be described by an abstract 'wave of probability', which spreads through space like a ripple on the surface of a pond. Wherever the wave is big, the probability of finding the particle is high; wherever it is small, the probability low. Once the particle has been 'observed', however, it ceases to be described by a haze of probabilities but instead becomes 100 per cent real. This explains why things in the real world do not appear in several places at once.

It turned out that there is a foolproof way of determining how the probability waves spread through space. It was discovered by the Austrian physicist Erwin Schrödinger in 1926, and was named the 'Schrödinger equation' in his honour. Applying it to, say, a hydrogen atom enabled physicists to understand at last what was so special about the stable atomic orbits whose existence had been inferred by Niels Bohr in 1913. They were simply the locations in the atom where it was most probable that the electron would be found.

The whole bizarre theory, in which the behaviour of microscopic particles was determined by peculiar waves of probability rippling through space, was known as 'quantum theory'. It was created in 1926 by a small band of young men, mostly in their early twenties, who had dared to break the chains of the old ways of thinking about atoms.

George Gamow, much to his disappointment, had been too young to join in the development of quantum theory. Nevertheless, he was one of the first people fully to grasp its implications. In fact, the reason he had headed straight for the University of Göttingen after completing his doctorate in Leningrad was because it had been the birthplace of many of the revolutionary new quantum ideas.

When Gamow arrived at the university, everyone was using the novel tools of quantum theory to understand how atoms were put together. Gamow, who had a strong aversion to crowded fields, decided to look elsewhere for problems to solve. He was immediately struck by the fact that nobody had thought of using the new theory to make sense of the nucleus which sat at the centre of atoms. So it was that Gamow had come to address the peculiar puzzle of alpha decay.

# THE HOUDINI-LIKE ABILITY OF WAVES

Since the behaviour of an alpha particle is determined by a probability wave spreading through space, in some sense it lives a schizophrenic existence as both a wave and a particle. Gamow's key insight was to realise that the wave-like character of an alpha particle might explain its extraordinary ability to escape from an atomic nucleus.

Waves can do all sorts of things that particles cannot. For instance, they can bend round corners, an ability which makes it possible to hear sound waves from someone who is out of sight. However, there is another trick that waves can perform, which Gamow realised had particular relevance to the process of alpha decay. The trick is easy to demonstrate for light waves which are travelling through a block of glass. If they strike the boundary of the glass at a shallow angle, they will not penetrate into the air beyond but will instead be reflected back into the glass. This process is known as 'total internal reflection'. Everything changes, however, if another block of glass is brought very close to the first. Now only some of the light is reflected internally. The rest leaps the air gap and crosses into the second piece of glass.

This phenomenon has its root in the fact that a wave is not localised like a particle, but is spread out through space. This means that light waves which are undergoing total internal reflection are not reflected from the precise boundary of the glass but penetrate into the air beyond for a distance of a few wavelengths. If these 'feelers' happen to encounter another block of glass before they can turn back, they generate new light so that the light continues on its way.

This ability of light to traverse a region of space apparently forbidden to it is common to every type of wave, and provided the key to understanding alpha decay. For if a light wave could leap out of one block of glass into another, reasoned Gamow, then the probability wave associated with an alpha particle must also be able to leak out of an atomic nucleus into the surrounding space. One moment, the alpha would be trapped inside a nucleus; the next it would appear as if by magic on the outside of its prison.

The extraordinary nature of the alpha's escape is evident in the volcano model of a nucleus. The trapped marble escapes, not by surmounting the lip of the crater, but by tunnelling its way through the walls of the volcano so that it emerges on the lower ramparts far below the summit. This is the reason why an alpha particle from uranium-235 has an energy of 4.2 MeV rather than of 17 MeV. Because of its wavelike nature, it is able to penetrate an apparently impenetrable barrier.

Inside a nucleus an alpha particle careers back and forth, battering itself against the walls of its prison thousands of billions of billions of times a second. And each time it strikes the walls there is a tiny chance of it tunnelling its way to freedom. How big that chance is depends on the size and shape of the barrier surrounding a nucleus and the depth of the volcanic crater.

If the crater is very deep, an alpha particle might never escape; this is the case for all stable atomic nuclei. If the crater is relatively shallow, a nucleus might decay extremely rapidly. In fact, the probability of an alpha particle escaping its prison is extremely sensitive to the depth of the crater – in other words, to the energy of the alpha particle. Making the crater even slightly deeper makes breaking free hugely more difficult. This is the reason that the half-lives of nuclei span such a fantastic range from the merest split-second to many billions of years.

Gamow wrote up his remarkable theory of alpha decay and sent it for publication while still at Göttingen. Coincidentally, a British physicist called Ronald Gurney and his American collaborator Edward Condon hit upon the tunnelling idea independently. Their paper appeared in print within a week of Gamow's.

Gamow, had he only known it, now had in his hands the means to answer Eddington's prayers and show how hydrogen might be turned into helium at millions rather than billions of degrees. Unfortunately, Gamow knew very little about stars, and so failed to recognise the significance of tunnelling for energy-generation in the sun. The man who did recognise it was another young physicist at Göttingen. His name was Fritz Houtermans.

## TURNING TUNNELLING ON ITS HEAD

Houtermans, like Gamow, was one of the more colourful characters of twentieth-century physics. Born in 1903 near the German port of Danzig, he grew up as an Austrian in Vienna. In the 1920s he joined the German communist party and was subsequently expelled from school for publicly reading its manifesto. Houtermans' family paid for the 'troublesome' teenager to be treated by Sigmund Freud, but Freud called a halt to the psychoanalysis when his patient admitted he had been making up his dreams. Houtermans' membership of the German communist party forced him to flee to England with his family when Hitler came to power in 1933. In London, working at His Master's Voice, Houtermans attempted to verify a prediction of Einstein's that the strength of a light beam would be boosted by passing it through a gas of 'excited' atoms. Had it not been for the failure of an expensive transformer which his employer refused to replace, he might well have invented the laser a quarter of a century before its time.

Houtermans' dislike of English cooking, and especially of the smell of boiled mutton wafting from cheap restaurants, caused him to accept a post at the Ukrainian Physico-Technical Institute in Kharkov. In the Soviet Union his naïve belief that communism offered a bright future for humanity was quickly dispelled. In 1937, at the height of Stalin's terror, his wife and two small children escaped to America. But he was trapped and arrested as a German spy by the notorious NKVD (the forerunner of the KGB). Three years of imprisonment, torture and near starvation ended only in 1940 when Hitler signed his notorious pact with Stalin.

Now in the hands of the Gestapo, Houtermans was spared yet more torture – this time as a communist spy – by the courageous intervention of the Nobel prizewinner Max von Laue, one of the few German scientists openly to defy the Nazis. Thereafter, Houtermans became one of the leading lights in the German nuclear programme, and the first to recognise the unique importance of plutonium for making an atomic bomb. Incredibly, he risked his life at the height of the war by sending a message to America's nuclear scientists: 'Hurry up. We are on the track.'

Despite all that had happened to him, however, Houtermans retained his sense of humour. In 1945, when every commodity was scarce in Germany, he actually persuaded the head of the Nazi nuclear project that heavy water, an essential component of a nuclear reactor, could be extracted from Macedonian tobacco. Before his game was rumbled, Houtermans had puffed his way through two whole bags of the precious stuff. Only the intervention of von Laue yet again saved him from the Gestapo.

In the summer of 1928 all these dramatic and terrible events belonged in the unknowable future. It was then that Houtermans first ran into George Gamow. The pair hit it off immediately. Houtermans was greatly impressed by Gamow's tunnelling theory. Shortly after Gamow published the idea, the two physicists worked on a joint paper in which they applied the theory to specific nuclei. However, by the time the account was ready for publication, Gamow's money had run out and he had gone to Copenhagen, while Houtermans had moved on to Berlin.

Houtermans continued to think about the alpha decay idea in Berlin, and was struck by an obvious thought which Gamow had overlooked. If particles could tunnel out of a nucleus, then they must also be able to tunnel in. Such a feat might be achieved by firing a subatomic projectile into an atom with sufficient violence that it drilled its way through the nuclear force barrier and lodged itself in the nucleus. It was an exciting prospect because it would open up a whole new class of element-building, or 'fusion', reactions. But was such a feat possible in the laboratory?

Disappointingly, Houtermans worked out that it was not. The subatomic projectiles available to him, chiefly alpha particles, simply did not have enough energy to make tunnelling into a nucleus at all likely, and he possessed no means to speed them up artificially.* However, it suddenly struck Houtermans that there was one place where higher energy particles were readily available: the centre of the sun.

## COOKING UP HELIUM IN A NUCLEAR POT

Houtermans knew very little about the sun. But, in Berlin, he got talking to a young British experimental physicist called Robert Atkinson. Through reading *The Internal Constitution of the Stars* Atkinson had learnt of the problem that the solar core was far too cold for the fusion of hydrogen into helium, the energy-generating process Eddington was convinced was powering the sun.

Houtermans, on hearing of this difficulty, was immediately struck by the thought that tunnelling would have a significant effect on any fusion processes going on in the sun. He interested Atkinson in exploring the idea and in the spring of 1929 the two men began their work. Their first task was to identify a way of turning hydrogen into helium.

In common with everyone else at the time, Houtermans and Atkinson believed that a helium nucleus was assembled from four protons. The obvious way for four protons to get together was for two protons to collide and stick, a third to bump into and join this pair, and a fourth to attach itself and complete the quartet. However, there was no known nucleus with two protons – helium-2 appeared not to exist. The obvious route was therefore blocked at the first step.

Forced to look around for a more indirect process, Houtermans and Atkinson came up with an ingenious scheme for making helium in which another nucleus helped things along by acting as a 'proton trap'. A proton might tunnel into such a nucleus, followed at some later time by another and another. Finally, after the capture of the fourth proton,

---

* Such a technique was later developed by two Cambridge physicists. In 1932 John Cockcroft and Ernest Walton bombarded nuclei with protons accelerated by a high voltage as had been suggested to them initially by George Gamow. For lodging a proton in a lithium nucleus, thereby turning it into a nucleus of the next heaviest element, beryllium, Cockcroft and Walton won the Nobel prize for Physics.

the nucleus would undergo a violent convulsion and cough out a fully formed alpha particle. By this means, hydrogen would be converted into helium, unleashing a flood of nuclear binding energy.

As Gamow had discovered, the tunnelling process was extremely sensitive to the energy of the tunnelling particles. In the sun the energy of those particles depended on the temperature. Any energy-generating process based on tunnelling would therefore have to be extremely sensitive to temperature – precisely the property Eddington had deduced it must have.

The details of the hydrogen to helium conversion were a little hazy to Houtermans and Atkinson. For instance, they did not know whether the binding energy would be liberated at once in the final, explosive ejection of the alpha particle, or whether it would somehow come out in dribs and drabs throughout the process. They also did not know the identity of the proton-trapping nucleus. However, it had to be capable of holding on to protons for a long time, be reasonably common in the sun, and be relatively light. Only a light nucleus, with a mere handful of protons, would have a repulsive barrier flimsy enough to be breached for tunnelling protons.

Houtermans and Atkinson now had a plausible scheme for turning hydrogen into helium. The next step was to determine whether, at the temperature Eddington maintained existed in the solar core, the scheme could generate enough energy to account for the heat output of the sun.

A simple calculation indicated that the average proton at the heart of the sun possessed barely a hundredth the energy necessary to tunnel into a typical light nucleus. However, some protons would be moving slower than average and some faster. And of those that were moving faster, some would be flying about at speeds many times greater than the average. Such ultra-fast protons would stand a far better chance of tunnelling into a light nucleus and making helium. The big question was therefore whether there were enough of them to create the heat of the sun.

Here the sun's enormous bulk was of crucial importance. For although only a tiny fraction of solar protons were moving fast enough to tunnel into a light nucleus and make helium, the sun contained a vast number of protons. A tiny fraction of a vast number might still be a huge number. Houtermans and Atkinson multiplied the relevant numbers together, and discovered that there were just enough ultra-fast protons for their purposes. Remarkably, the proton-capture process, driven by only the rarest, high-speed protons inside the sun, was capable of accounting for sunlight.

It was the answer to Eddington's prayers. Houtermans and Atkinson, by turning Gamow's tunnelling idea on its head, had shown that nuclear energy could power the sun at a temperature almost a thousand times lower than anyone had thought possible. That night, as the physicist Robert Jungk relates in his book *Brighter than a Thousand Suns*, Houtermans tried to impress his girlfriend with a line that nobody in history had ever used before. As she stood entranced by a perfect moonless sky, he boasted he was the only person in the world who knew why the stars were shining. It apparently worked. Two years later, Charlotte Riefenstahl agreed to marry him! (In fact, she would marry him twice – the second time in 1953, after their enforced wartime separation.)

Houtermans and Atkinson had demonstrated that at the temperature found inside stars nuclei could penetrate into each other and cause nuclear reactions, releasing energy. What was so remarkable about their proof was that they had obtained it despite an overwhelming ignorance of nuclear physics! In fact, it is fair to say that the science of nuclear physics did not exist before James Chadwick discovered the neutron in 1932.[*] For how could anyone say anything sensible about the atomic nucleus while unaware of one of its two main constituents?

The ignorance of the neutron's existence meant that in 1929 Houtermans and Atkinson, along with everyone else, believed that a helium nucleus was assembled from four protons. Three years later, with the realisation that a helium nucleus in fact contained two protons and two neutrons, the proton-trapping scheme for powering the sun was confronted by a very serious problem.

An obvious solution was to assume that the trapping nucleus simply gobbled up two protons and two neutrons. Unfortunately, there is no convenient source of free neutrons. Although the particles can exist indefinitely inside a nucleus, on the outside they self-destruct within a matter of minutes. The only possible way to salvage Houtermans and Atkinson's scheme was to postulate that, inside the trapping nucleus, two of the four gobbled-up protons somehow changed themselves into neutrons. But was such a bizarre transformation possible?

The answer was a long time in coming. But, in 1938, a process by which a nuclear proton might metamorphose into a neutron was suggested by the physicist Carl-Friedrich von Weizsäcker, the son of the second highest official in Hitler's foreign ministry. The proton

[*] Some would date the birth of nuclear physics to 1934, by which time it was universally accepted that the neutron was a particle in its own right, and not merely a proton and an electron sandwiched together.

accomplished the feat by emitting two particles – a positron and a neutrino.[*]

The time was now ripe to flesh out Houtermans and Atkinson's vague scheme, and to determine the precise sequence of nuclear reactions that was powering the sun. Many of the big guns of theoretical physics, including Robert Oppenheimer, the 'father' of the atomic bomb, had tried and failed to find the elusive recipe for sunlight. Success would go to two men: von Weizsäcker and an American physicist called Hans Bethe.

[*] Positrons are positively charged electrons; they had been discovered in 1931. Neutrinos are ghostly particles which pass through matter as if it was not there; predicted by theory, they would not be detected experimentally until the 1950s.

# 8

# A Recipe for Sunlight

## HOW WE DISCOVERED THE PRECISE MANNER IN WHICH HYDROGEN IS CONVERTED INTO HELIUM INSIDE STARS.

*It should not be so difficult after all to find the reaction which would just fit our old sun. I must surely be able to figure it out before dinner!*

Hans Bethe (as remembered by George Gamow)

Hans Bethe's mother was Jewish. In 1933, therefore, when Hitler came to power, the 27-year-old physicist was instantly dismissed from his post of assistant professor at the University of Tübingen. The man who fired him, apparently without a twinge of regret, was Hans Geiger, whose alpha-scattering experiments had revealed the atomic nucleus to Rutherford.

Bethe, like Fritz Houtermans, fled to England. However, unlike his fellow countryman, he had the good fortune to obtain a post in the United States, rather than the Soviet Union. At New York's Cornell University Bethe quickly established himself as a central figure in the rapidly advancing field of nuclear physics, writing a series of reviews of the subject which were so influential they were referred to as 'Bethe's Bible'. His mastery of nuclear physics would ensure that, when war with Germany came, he was chosen to head the theory division at Los Alamos the secret laboratory charged with developing the atomic bomb.

Bethe's interest in nuclear physics had been a down-to-earth one. He was aware of the problem of stellar energy generation, but considered the task of identifying the nuclear reactions powering the sun too speculative to waste time on. His opinion changed, however, after attending a conference in Washington DC in the spring of 1938. He would have ignored the conference had it not been for his friend, the Hungarian physicist Edward Teller. It was Teller, a fellow refugee from Europe and the man who would one day 'father' the American hydrogen bomb, who urged him to go. What clinched it for Bethe was that the conference was being organised by George Gamow.

Gamow, after wandering throughout Europe, had arrived in the United States in 1934 to take up a post at George Washington University. Like Bethe, he was active in down-to-earth nuclear physics. However, he had also developed a keen interest in stars, chiefly because Houtermans and Atkinson had shown his tunnelling idea to be crucial to the understanding of any nuclear reactions in the sun. It was this interest that had prompted him to organise a three-day meeting on the 'stellar energy problem'.

Bethe had low expectations for the conference. However, as he sat listening to the astronomers describing the latest view of the conditions in stars, it struck him suddenly that he had been wrong about the stellar energy problem. With his command of nuclear physics, there was a very good chance that he would be able to identify the elusive nuclear reactions that powered the sun.

According to Gamow, it was on the train journey home to New York that Bethe discovered the reactions responsible for sunlight. Intending to solve the problem before dinner, he took out a sheet of paper and began covering it with formulae and tables of numbers. One after another he discarded possible nuclear reactions, a task in which he was still engaged when the sun dipped below the horizon and the serving of dinner on the train became imminent. 'But Hans Bethe was not a man to miss a good meal simply because of some difficulties with the sun,' said Gamow. 'Redoubling his efforts, he found the correct answer at the very moment when the passing dinner-car steward announced the call for dinner!'

It would be wonderful if scientific discoveries were really made in such a romantic manner. Nevertheless, all good stories contain some truth. And it is certainly true that, within six weeks of the conference, Bethe was telling everyone he had solved the stellar energy problem.

## THE CARBON CYCLE

The first difficulty Bethe faced was to determine the identity of the elusive proton-trapping nucleus. As Houtermans and Atkinson had realised, it would have to be light, because the repulsive barrier around a heavy nucleus would be too formidable to be penetrated by a tunnelling proton.

But the trapping nucleus, in addition to being light, would also have to possess another important property: it had to take hold quickly of any proton that happened to tunnel through to meet it. If it reacted too slowly, the proton would tunnel back out again and escape. It was because Houtermans and Atkinson had absolutely no idea of the speed of nuclear reactions – which nuclei would react quickly with protons and which slowly – that they were unable to identify a suitable proton-trapping nucleus.

However, by the late 1930s physicists knew enough about nuclear reactions for Bethe to be able to work his way systematically through a list of the lightest nuclei, determining in each case how enthusiastically it would snap up a proton.

Helium, the second lightest element, turned out to be useless. There was simply no stable nucleus of mass 5 – that is, with a mass five times that of a hydrogen atom. Next, Bethe tried lithium, beryllium and boron, all to no avail. In fact, they gobbled up protons too fast; these elements would be used up too rapidly inside a star.* Then Bethe came to element number six. To his delight, he found that it reacted very swiftly with protons. His quest was over. He had found the elusive proton-trapping nucleus, the 'pot' in which hydrogen might be cooked into helium. It was carbon.

Next, Bethe began to determine the sequence of nuclear reactions by which a carbon nucleus would turn hydrogen into helium. The reaction chain he came up with started with a proton tunnelling into carbon-12 and finished, after three more protons had followed suit, with the splitting of the final nucleus into a nucleus of carbon-12 and a nucleus of helium-4. The helium-4 was then ejected as an alpha particle.

The details of the process were complex. After the capture of each proton, the trapping nucleus became unstable and sooner or later disintegrated. These decays accomplished two things. Firstly, they turned two of the trapped protons into neutrons in the way suggested by Carl-Friedrich von Weizsäcker. And secondly, they spat out gamma rays, positrons and neutrinos. It was by these means that nuclear binding energy was converted into other forms. Ultimately, of course, these other forms of energy would be turned into heat: the heat of the sun.

The proton-trapping nucleus actually went through several incarnations. The carbon-12 nucleus first changed into nitrogen-13, then into carbon-13, nitrogen-14, oxygen-15 and nitrogen-15. Finally, after

---

* This is precisely why lithium, beryllium and boron are rare on earth.

spitting out the alpha particle, it returned to being carbon-12. The regeneration of the carbon at the end of the process meant it could be used again. This was extremely convenient because carbon was rare in stars. The only raw material that Bethe's process in fact consumed was hydrogen, the stars' most abundant element. Because the sequence of nuclear reactions returned to its starting point, it was more a cycle than a chain. It therefore became known as the 'carbon cycle'.*

The carbon cycle turned out to be astonishingly sensitive to temperature. The energy it generated grew as the 24th power of the temperature, which meant that doubling the temperature of a star would increase its energy production by more than 10 million times.†

At the time Bethe solved it, the stellar energy problem was ripe for the solving. In Germany, von Weizsäcker discovered the carbon cycle entirely independently. In fact, the two men submitted their papers to journals within a few months of each other. Both had hit on carbon as the elusive proton-trapping nucleus after systematically examining the nuclear reactions between hydrogen and common light nuclei. However, the lightest nucleus of all was hydrogen. What about nuclear reactions between hydrogen and hydrogen – in other words, ones in which hydrogen itself played the part of the proton-trapping nucleus?

## THE PROTON-PROTON CHAIN

Gluing together protons in this way was of course the obvious way to make helium. It had been dismissed by Houtermans and Atkinson because they knew of no nucleus in nature consisting of two protons. However, everything was changed by a discovery made just after Chadwick's announcement of the neutron. In 1932, the American chemist Harold Urey found a heavy isotope of hydrogen. It contained a single proton and a single neutron and was christened deuterium.‡

Deuterium, by virtue of its two proton-like particles, was the nearest thing in nature to a nucleus with two protons. This prompted an intriguing thought. Was there any way in which two protons might collide in the sun and make deuterium? If so, it would open up the

---

* It is also called the carbon-nitrogen-oxygen cycle, or the CNO cycle for short.

† The enormous dependence on temperature of the nuclear reactions inside stars is due to the tunnelling process, which is extraordinarily dependent on the velocity of the tunnelling particles. Their velocity, in turn, is determined by the temperature in the star.

‡ Deuterium is present in nature in rather small amounts and one in every 7000 molecules of water contains an atom of deuterium. Such water molecules are known as 'heavy water'.

straightforward route to forging helium. Of course, one of the protons would have to change into a neutron. However, von Weizsäcker had suggested how such a transformation might occur, and both he and Bethe had incorporated such a process in their carbon cycle.

The formation of a deuterium nucleus, accompanied by the release of a burst of energy, was merely the first step towards making a helium nucleus. In Germany, at essentially the same time he had worked out the carbon cycle, von Weizsäcker had worked out the details of how such a conversion might occur. In the United States, Bethe did the same, working with Charles Critchfield, a student of Gamow and Teller's whom he had met at the stellar-energy conference in Washington DC.

In the straightforward helium-building process, which became known as the proton-proton chain, a proton tunnelled into a deuterium nucleus to make a nucleus of helium-3, a light isotope of helium, releasing more energy. Then, in a departure from proton tunnelling, a helium-3 nucleus actually tunnelled into another helium-3 nucleus, unleashing yet more energy. This created the desired nucleus of helium-4, with two protons left over to start the process again.*

Now, with the proton-proton chain and the carbon cycle, there were two recipes for sunlight. But which one was powering the sun? According to Bethe's calculations, both appeared capable of generating about the right amount of energy. However, his calculations were not precise enough to favour one over the other.

The protons involved in the carbon cycle were faced with a far more formidable repulsive barrier than those involved in the proton-proton chain. They would therefore have to be moving faster to succeed in tunnelling. On the other hand, once a proton had tunnelled into a carbon nucleus, it was far more likely to react with it than a proton with a proton. All that could be said with any certainty was that, at low temperatures, where protons were moving too slowly to tunnel into carbon, the dominant energy-generating process in stars would be the proton-proton chain. At high temperatures, it would be the carbon cycle.

In order to decide which process was dominant in the sun, it was necessary to determine how quickly the individual nuclear reactions of the two cycles took place. But no theorist, not even Bethe, possessed the mathematical tools to predict these speeds. They would have to be measured in an experiment. Such an experiment would be carried out by a young American nuclear physicist named Willy Fowler.

---

* In fact, there are competing reactions, in which a helium-3 nucleus combines with a helium-4 nucleus to yield a nucleus of beryllium. This then turns into two helium nuclei, either by decaying radioactively or by capturing a proton.

# A RECIPE FOR SUNLIGHT

Fowler was a stocky extrovert from southern Ohio. In 1933, he arrived at the California Institute of Technology in Pasadena, to begin work in the Kellogg Radiation Laboratory. The laboratory, built with money from the Detroit cornflake magnate W. K. Kellogg, had been built two years earlier as a high-energy X-ray facility, with the X-rays used for treating cancer patients in the day and for physics experimentation at night. Everything changed, however, in 1932 with the dramatic news from England that John Cockcroft and Ernest Walton had 'split the atom' with artificially accelerated protons. Immediately, Kellogg's director, Charles Lauritsen, set about converting one of the lab's million-volt X-ray tubes to accelerate protons and other nuclei, rather than electrons. Now in possession of an American version of the British 'atom smasher', Lauritsen looked for young experimentalists to come to Kellogg and work as his assistants. One of the men he recruited was Willy Fowler.

Lauritsen and Fowler concentrated on bombarding light nuclei such as carbon and nitrogen with high-velocity protons, to see what nuclear reactions were triggered and how fast those reactions went. It was in 1934, during the course of this work, that Lauritsen discovered the peculiar ability of carbon-12 to trap a proton, rather than simply to disintegrate when struck. In the process, the nucleus was transformed into nitrogen-13.

At the time of the discovery, neither Lauritsen nor Fowler had any idea of its wider implications. However, the pivotal moment in Fowler's career came in the spring of 1938, during a conversation with the theorist Robert Oppenheimer. Oppenheimer, who was dividing his time between the University of California at Berkeley and Caltech, happened to mention Hans Bethe's idea that a process called the carbon cycle, which turned hydrogen into helium, might generate the energy of stars. It began when a nucleus of carbon-12 captured a proton to yield a nucleus of nitrogen-13. Fowler was stunned. It was the very reaction he and Lauritsen had already observed. What everyone had been doing in the depths of Kellogg's basement had greater importance than any of them had ever imagined.

Fowler found a copy of Bethe's 'Physical Review' paper on the carbon cycle in Caltech's library. He read it several times with mounting excitement. When he put it down, the seed of an idea had been planted in his mind. In Kellogg's laboratory he would reproduce every single step of Bethe's recipe for sunlight.

The Second World War interrupted Fowler's plans, and it was not until 1946 that he and his student Bob Hall embarked on a programme to measure the speeds of the individual nuclear reactions in the carbon cycle. The experiments were slow and very difficult but, by the early 1950s, Fowler and Hall had come to a definite conclusion. The carbon cycle could not be the dominant energy-generating process in the sun. The temperature in the solar core was a few million degrees too low. Fowler's measurements quickly confirmed the only alternative explanation. The driving force behind sunlight had to be the proton-proton chain.

## A GIANT CONTROLLED HYDROGEN BOMB

It was official: the sun was a giant hydrogen bomb. However, it differed in two crucial respects from a man-made nuclear weapon.

First, the reactions going on in the solar interior are as far from explosive as it is possible to imagine. It takes each proton in the core of the sun an astonishing 10 billion years to find and react with another proton to make deuterium.* Subsequent reactions in the proton-proton chain are rapid by comparison. However, it is the unbelievable slowness of the first step which is all-important to the generation of solar heat. It ensures that the sun consumes its fuel very gradually over many billions of years.†

In fact, the reaction between two protons to make a nucleus of deuterium is so slow that even the human body generates more heat, volume for volume, than the sun! The reason the sun is so hot, despite such a pathetically low rate of heat production, is simply that it is large. Like all large bodies, its surface area is small compared to its volume. It means that solar heat, which can escape only through its surface, gets dammed up inside.

The second important respect in which the sun differs from a

---

* The technical reason for the slowness of the reaction has to do with the fact that there are really two nuclear forces – the 'strong' and 'weak' nuclear force. The weak force alone can change a neutron into a proton (or vice versa), and so is responsible for both radioactive beta decay and for making deuterium in stars. However, the weak force is very slow to act compared with the strong force. For this reason, two protons which meet face to face after penetration invariably separate again long before the weak force can transform them into deuterium.

† If it were possible for two protons to bind together to make helium-2, it would happen very quickly, and all the universe's hydrogen would long ago have been turned into helium-2, leaving no hydrogen to burn inside stars. The strong nuclear force is in fact only very slightly too weak to bind helium-2; our very existence hinges on an extremely fortunate accident of nuclear physics.

hydrogen bomb is in possessing an in-build 'safety valve', which prevents runaway nuclear reactions. If for some reason the fusion reactions in the core speeded up, the extra energy produced would make the sun expand slightly. However, such an expansion would lower the temperature in the core, slowing the energy-generating fusion reactions and curing the original overproduction of nuclear energy. On the other hand, if the fusion reactions in the core slowed down, the energy deficit would cause the sun to shrink. The shrinkage would raise the temperature of the core, boosting the rate of energy generation and once more curing the original problem.

Fortunately the sun is a very controlled hydrogen bomb, and the source of its heat – the proton–proton chain – is just about the most inefficient nuclear reaction conceivable.

Two and a half millennia after Anaxagoras had guessed that the sun is 'a red-hot ball of iron not much bigger than Greece', science had finally provided the definitive answer to the question: what makes the sun shine? In fact, the final proof that the proton–proton chain is indeed the ultimate power source of the sun had to wait until 1988 when an experiment in the Kamioka metal mine, in the Japanese Alps, detected neutrinos coming from the direction of the sun. These ghostly particles, which are rarely stopped by matter, are an unavoidable by-product of the proton–proton chain.

But although Fowler's measurements showed that the proton–proton chain is responsible for sunlight, they did not consign the carbon cycle to the dustbin of scientific history. The measurements indicated instead that Bethe and von Weizsäcker's cycle for turning hydrogen into helium is the dominant source of energy in stars which are hotter, and consequently more massive, than the sun.

But the carbon cycle posed a conundrum. In order to work, it required a star to contain a supply of carbon. But where did the carbon come from? Where, for that matter, did all the elements heavier than helium come from? What was the origin of the iron in our blood, the calcium in our bones, the oxygen that fills our lungs each time we take a breath? An obvious possibility was that these elements had always been in existence – that they had simply been placed in the universe by the Creator on Day One. However, evidence had been mounting for some time that this was not the case, that helium was not the only element that had been made. The vital clue had been found by Francis Aston.

Part Three

# The Magic Furnace

# 9

# *The Forge of God*

HOW WE DISCOVERED THAT THE MIX OF
ELEMENTS IS THE SAME EVERYWHERE
IN THE UNIVERSE, AND REALISED
THAT THE ATOMS HAD ACTUALLY
BEEN MADE.

> Twinkle, twinkle little star
> I don't wonder what you are,
> For by spectroscopic ken,
> I know that you are hydrogen.
>
> Ian Bush

After his discovery of the anomalously high mass of hydrogen, Francis Aston greatly improved his mass spectrograph, and began using it to make increasingly precise measurements of the weights of atoms. In 1925, this perseverance paid off when he made another important discovery. Hydrogen, with its weight of 1.008, was not alone in having a mass that differed from a whole number. So too did almost all other atoms.

In every case the discrepancy was much smaller than for hydrogen. The mass of helium turned out to be 4.003 rather than 4, the mass of oxygen 15.995 rather than 16.* There was a pattern, but it became apparent only after Aston had accumulated scores of similar measurements.

Broadly speaking, atoms fell into two categories: those whose weight was less than the nearest whole number and those whose weight was

---

* Once again, these are numbers Aston would have obtained if he had compared his atomic masses with carbon-12, as chemists do today.

greater than the nearest whole number. In the first category were atoms of medium weight, like oxygen at 15.995 and chlorine at 34.969. In the second category were light atoms, like helium at 4.003, and heavy atoms, like radon at 226.024 and thorium at 232.038.

The nearest whole number was simply the number of hydrogen building blocks in a nucleus – helium had 4, oxygen 16, and so on.[*] Aston had therefore discovered that the nuclei of medium-weight and light atoms had the least energy per nuclear building block, while the nuclei of light atoms and heavy atoms had the most energy per nuclear building block. 'Least energy' is the same as saying 'most tightly bound'. Aston's results therefore revealed that the nuclei of medium-weight atoms are more tightly bound than the nuclei of light and heavy atoms.

Finding a variation in the binding of nuclei was not surprising. After all, the stability of a nucleus depends on the balance between two forces – the electrical repulsion trying to blow it apart, and the nuclear force trying to keep it together – and that balance was sure to be different for nuclei made of different numbers of building blocks. The surprise was that the energy per nuclear building block varied in such a systematic fashion.

Plotted on a piece of graph paper, it looked like a valley. High on one side of the valley was hydrogen, the lightest element. The slope then fell away steeply as nuclei became progressively heavier and heavier, finally bottoming out at iron-56, the lowest-energy, most tightly bound and so most stable nucleus in nature. Beyond iron-56, the slope rose again towards thorium and uranium, the heaviest elements, perched high on the far side of the valley.

The 'curve of binding energy', as it became known, contained a wealth of meaning. For a start, it illuminated one of the central mysteries of radioactivity: why the phenomenon is largely a property of heavy atoms like radium and uranium. The most stable state of any system is usually the one with the least possible energy. For a football in a valley, for instance, the most stable state is with the football at the bottom of the valley, minimising its gravitational energy. Given the slightest chance, gravity will roll the football down the side of the valley to this lowest-energy state.

And what is true for footballs experiencing the force of gravity is equally true for atomic nuclei feeling the forces that dominate their existence. Given the slightest chance, the nuclear forces will drive nuclei down to the lowest-energy state – that is, to the bottom of the curve of binding energy. For a heavy nucleus, perched high on one side of the

---

[*] After the discovery of the neutron, it would be recognised to be the number of protons plus neutrons in a nucleus.

valley, the down direction is towards lighter nuclei. This means that heavy nuclei can get nearer the valley bottom by disintegrating into lighter nuclei – that is, by undergoing radioactive decay.

Circumstances are very different for light nuclei, high on the other side of the valley, and medium-weight nuclei, in the valley bottom itself. If any of these were to change into lighter nuclei, they would have to move farther away from the valley bottom, which is as unlikely as a ball rolling uphill. Consequently, common isotopes of light and medium-weight atoms are not radioactive.

But this is by no means all that can be deduced from the curve of binding energy. For a light nucleus, the down direction is in the direction of heavier nuclei. Light nuclei can therefore get nearer the valley bottom, lowering their energy per nuclear building block, by building themselves into heavier nuclei. A perfect example of this 'fusion' process is of course the conversion of the lightest nucleus – hydrogen – into the second lightest nucleus – helium – where the surplus energy ends up as sunlight.[*]

Since the slope of the curve of binding energy is downhill all the way to iron-56, it means that in principle fusion could continue beyond helium, creating bigger and bigger nuclei. The proviso is that two nuclei get close enough to feel the nuclear glue and stick. In practice, this requires them to be slammed together violently enough to overcome their mutual electrical repulsion, which requires a lot of heat.[†]

The message in Aston's curve of binding energy is therefore that light nuclei can reduce their energy by fusing to make heavier nuclei and that heavy nuclei can reduce their energy by disintegrating into lighter nuclei. As with all sweeping statements, however, there are exceptions.

Helium, for instance, has less energy per building block, and is much more strongly bound, than nuclei which are slightly heavier. In fact, it is so tightly glued together that it retains its integrity even inside a large nucleus. This is the reason why, when radioactive nuclei hiccup, they spit out fully formed helium nuclei – alpha particles.

Helium's low energy creates a pothole on the side of the curve of binding energy. And it is not the only pothole on the slope as it sweeps down the valley. There is one at carbon-12. One at nitrogen-15. Another at oxygen-16. And the biggest one, of course, is the valley bottom around iron-56. The fact that carbon, nitrogen, oxygen and iron

---

[*] In fact, the proton-proton process is possible only because every step in the process is downhill. Deuterium has less energy than hydrogen; helium-3 has less than deuterium; and helium-4 has less than helium-3.

[†] Unless, of course, the nuclei are fired at each other artificially – the approach of Ernest Rutherford, who made oxygen-17 by firing alpha particles at nitrogen-14.

all correspond to potholes on the curve of binding energy may not seem very significant. However, these nuclei also have something else in common which makes the coincidence very significant indeed. This became apparent as soon as people learnt to estimate which elements were common in the universe and which were rare.


## A UNIVERSAL MIX OF ELEMENTS

When William Huggins had demonstrated that the same atoms on earth were also present in the stars, it had been a tremendous discovery. However, neither Huggins or anyone else had any way of determining which of those elements were common in the universe and which were rare. It soon became clear that there were wide variations in the spectra of stars. The obvious assumption to make was that differences in stellar spectra reflected differences in stellar compositions – for instance, that stars with prominent dark lines due only to hydrogen contained only hydrogen, and that stars with prominent lines due only to heavy elements contained only heavy elements.

When this kind of reasoning was applied to the sun, whose spectrum was filled with lines of terrestrial elements, it led to the idea that the sun was not only made of the same elements as the earth, but that those elements were present in the sun in the same proportions as on earth.

The truth, unfortunately, was that there was not the slightest evidence that the elements with the most prominent spectral lines in a star's spectrum were also the most common elements in the star. It was impossible to deduce anything meaningful about the abundance of elements without an understanding of what actually caused spectral lines. And such an understanding was out of the question before both the electron and the atomic nucleus were discovered, and Niels Bohr put these two pieces together in 1913 to form the first plausible picture of the atom.

In Bohr's atom, electrons were restricted to certain orbits around a nucleus, and spectral lines arose because an electron, jumping from one orbit to another, emitted or absorbed light of a precise wavelength. That wavelength corresponded to the difference in energy of the orbits. One important consequence of Bohr's atom was that if an electron received sufficient energy it might be catapulted clean out of an atom. As Eddington alone had recognised, this meant that what lay in the blisteringly hot interior of the sun was a restless plasma of charged nuclei and ejected electrons – a great simplification which opened the door to an understanding of stars.

The fact that electrons could be ejected from atoms also had important implications for the surface of the sun, for it turned out that the spectral lines from an atom bereft of one or more of its electrons were different from those from the same atom with all of its electrons. Suddenly, there was the possibility of a host of new spectral lines from each element.

In fact, astronomers had long known that the solar spectrum contained a forest of unidentified dark lines, which completely outnumbered the spectral lines of familiar elements. The mystery was partly solved in the 1860s when Norman Lockyer, the discoverer of helium, found that the spectral fingerprint of an element changed when it was heated to a very high temperature. He concluded that the unknown lines in the sun were produced by known elements but in a superheated state.

Lockyer speculated that spectral changes were caused when extreme heat caused atoms to disintegrate into their subatomic building blocks, which he believed to be the hydrogen atoms of William Prout. After Bohr, however, it was clear that such atomic disintegration instead involved the loss of electrons. In the searing-hot gas of a star, violent collisions chipped these tiny particles from atoms.

The consequence of this for an element such as iron was that at any one time it might exist in several different 'ionisation states': some iron atoms might be neutral, others stripped of a single electron, still others of two electrons, and so on. The average number of electrons lost by an iron atom depended largely on the temperature, since the hotter the gas the more violent were the collisions responsible for removing electrons.* It followed that if a star was extremely hot, almost all its iron atoms would have lost some electrons. With very few in the neutral state – that is, with their full complement of electrons – the star's spectrum might very well contain no discernible lines due to neutral iron. So just because a star's spectrum showed no sign of neutral iron, it did not necessarily mean that the star contained no iron. The star might contain plenty of iron but be so hot that hardly any of its iron atoms were in the right state to announce their presence.

This was the crucial insight for decoding the message in starlight. The elements that made their mark in a star's spectrum were not necessarily the elements that were most common in the star. They were merely the ones that happened to stand out at the particular temperature of the star. The differences between the spectra of stars were not so much due to

---

* In fact, the prevalence of free electrons also played a role, since the more free electrons there were around bumping into atoms the likelier it was that yet more electrons would be lost.

differences in their composition but to differences in their temperatures.[*]

Stars, it now appeared, were remarkably uniform. Not only did they contain the same elements as the sun and earth, but they contained them in roughly the same proportions. Remarkably, there was a universal mix of elements.

But what was that universal mix? Despite their progress in understanding spectra, astronomers were still no nearer to deducing from a stellar spectrum which elements were common and which were scarce. What was needed was a way of disentangling the effect of a star's composition on its spectrum from the much larger effect of its temperature.

The breakthrough came in 1920 when a young Indian physicist called Meghnad Saha successfully combined Bohr's picture of the atom with the theory of heat. For any given temperature and density, the Saha equation predicted what fraction of an element's atoms would be neutral, stripped of one electron, of two electrons, and so on. It was a crucial step towards being able to determine how many atoms of an element were needed to generate a stellar spectral line. The final step was taken in 1923 by the English physicists Arthur Milne and Ralph Fowler. However, they did not apply their method of estimating the abundances of elements to real stars. That was left to an Englishwoman called Cecilia Payne.

## THE SUPER-ABUNDANCE OF HYDROGEN AND HELIUM

Cecilia Payne had become fascinated by the problem of decoding the message in starlight while still an undergraduate at Cambridge. However, she was forced to go to America to pursue the problem because of the exclusion of women from British astronomy. Payne enrolled for a doctorate at Radcliffe College, which was on the doorstep of Harvard Observatory, the home of the world's richest collection of stellar spectra. She quickly set about applying Milne and Fowler's method to some of the Harvard spectra, obtaining her first results in 1925. Those results contained a bombshell.

Some of the most prominent lines in the solar spectrum were due to the lightest element, hydrogen. This was unexpected, because Payne's

---

[*] Stellar temperatures range from about 3,000°C for cool, red stars to about 50,000°C for blisteringly hot, blue-white stars.

calculations indicated that, at the temperature of the sun's surface, only a minuscule fraction of all hydrogen atoms should be in the necessary state to produce hydrogen lines. The only reasonable explanation was that this fraction must still represent a very substantial number of atoms. In other words, hydrogen atoms must be extraordinarily abundant on the sun. Similar logic applied to the second lightest element, helium. In fact, Payne's calculations seemed to be implying that the two lightest elements made up an astonishing 98 per cent of the mass of the sun!

It was too incredible to be true. In her doctoral thesis, published in 1925, Payne called the result 'spurious'. She even went so far, in a scientific journal, as to declare, 'The abundance of both hydrogen and helium in stars is improbably high and is almost certainly not real.' In rejecting what would one day go down as her greatest discovery, Payne was heavily influenced by Henry Norris Russell, the American astronomer who had discovered red giants. Russell pointed out that Payne's results showed that all other elements, such as silicon and carbon and oxygen, existed in the sun in the same relative proportions as they did in the earth's crust. Why, if the sun and the earth were so similar for most elements, should they differ so much for hydrogen and helium?

Such an argument by analogy was a weak one, but Russell was one of the most influential astronomers in the world, and his criticism carried a lot of weight. Nevertheless, Russell was eventually forced to admit he had made a mistake. By 1929, the evidence for the super-abundance of hydrogen, in particular, was overwhelming. And if hydrogen was super-abundant on the sun, it was likely to be super-abundant *in* the sun as well. There could no longer be any doubt that the sun had sufficient hydrogen fuel to keep it burning for billions of years.

A hydrogen sun also resolved a major problem with Eddington's theory. This predicted a light output from the sun that was a hundred times bigger than observed. Eddington had assumed that the sun was made mostly of iron. However, if it was instead made of hydrogen, it would have to contain more particles, because hydrogen atoms are much lighter than iron atoms.[*] Since the pressure of gas goes up if there are more particles hammering away, hydrogen would be more efficient at generating pressure in the solar core. It could support the weight of the outer layers of the sun with a lower central temperature: 13 million degrees, rather than the 40 million degrees calculated by Eddington. And a lower central temperature meant a much lower solar light output.[†]

---

[*] A hydrogen plasma actually contains about 4 times as many particles – nuclei plus protons – as an iron plasma of the same density.

[†] Eddington's theory could reproduce the light output of the sun with two distinct

If helium is also very abundant in the sun, this would explain why it had been so easy for Norman Lockyer to spot it in a solar prominence in 1868. There was no need to postulate that the element was a by-product of radioactivity, as Rutherford had done. Helium's fingerprint was visible in the solar spectrum simply because the element was a major constituent of the sun.

The discovery by Payne that the sun contains huge quantities of the two lightest elements had implications far beyond the sun. Since the stars were known to have roughly the same mix of elements as the sun, they too must be made mostly of hydrogen and helium. Nobody had ever suspected this, and with good reason: hydrogen and helium are almost non-existent on earth. It was now clear, however, that the absence of hydrogen and helium on earth has nothing whatsoever to do with their cosmic abundance and everything to do with their extreme lightness. Any hydrogen and helium present on the earth when it formed has long ago floated off into space.

Payne's estimates of the element abundances in the sun therefore indicated that hydrogen and helium were the most common elements in the entire universe. Hydrogen accounts for almost 90 per cent of the atoms in the universe, with helium making up most of the remaining 10 per cent. All the other elements are no more than a minor contaminant of the cosmos, accounting for little more than 2 per cent of its mass. Put more precisely, the evidence from the stars is that for every 10,000 atoms of hydrogen in the universe, there are 975 atoms of helium, 6 atoms of the third most common element, oxygen, and only 1 of the fourth most common, carbon. No other element reaches a level of even 1 atom per 10,000 hydrogen atoms, and the vast majority fall a long way short of this.

But estimating the cosmic abundances of the elements simply from how common they are in stars is a risky business, since stellar spectra are always difficult to interpret. Improvements could be made by supplementing the estimates from stars with estimates from objects that could be examined – for instance, the rocks of the earth's crust, and meteorites that have fallen to earth. The first person to do this was a Swiss-Norwegian chemist called Victor Goldschmidt in 1936. He and his successors uncovered a suggestive pattern in the relative abundances – the abundance curve – of the different elements.

------

recipes – one with 35 per cent hydrogen and the other with 90 per cent hydrogen. The higher (correct) figure proved difficult for astronomers to believe and so Anaxagoras' idea of an iron sun survived almost to the 1950s.

Leaving aside the two lightest elements, hydrogen and helium, elements became rapidly less common as they got heavier and heavier. The decline in abundance was so steep that a heavy element like silver, with an atomic mass of 108, was a millionth as abundant as a light element such as magnesium, with a mass of 24. Plotted on a graph, it looked like a sheer mountainside. However, there are notable exceptions to the rapid decline in abundance with growing atomic weight. Some elements are distinctly more common than their nighbours, creating bumps on the mountainside, such as carbon, nitrogen, oxygen, and elements in the vicinity of iron. Other elements are distinctly less common than their neighbours, creating hollows, such as lithium, beryllium and boron.

Oddly, the bumps on the abundance curve coincided with the potholes on the curve of binding energy. And the hollows on the abundance curve coincided with the small hillocks on the curve of binding energy. It was as if a number of footballs had rolled down a slope, avoiding the bumps and becoming trapped in the potholes. What did it mean?

Lithium, beryllium and boron are more loosely bound than neighbouring nuclei, while carbon, nitrogen, oxygen and iron are more tightly bound. So the most weakly bound nuclei were rarer than expected, and the most tightly bound were more common. The abundances of each element clearly reflect the properties of its nuclei. There could be only one explanation. Nuclear processes must have been involved in *creating* those abundances. Here was the first compelling evidence that all elements – not just helium – had been made.

But how had this happened? What universal process had forged the elements?

## A RECIPE FOR ELEMENT-BUILDING

Ever since William Prout, the idea had been in the air that all atoms were assembled from a simple building block. In Prout's day the obvious candidate was the hydrogen atom. With the discovery of the atomic nucleus, however, attention shifted to the nucleus of the hydrogen atom, the proton.

In the 1930s, Hans Bethe, Charles Critchfield and Carl-Friedrich von Weizsäcker established that stars like the sun live off the tremendous binding energy liberated when protons fuse together to make helium nuclei. With this precedent in mind, it was natural for physicists to imagine making heavier and heavier atoms by simply sticking together more and more protons. This was a downhill process – at least until iron-56 – and so would liberate more and more nuclear binding energy.

The trouble is that such a proton build-up process faces an insurmountable barrier: the tremendous repulsion between an incoming proton and the protons in a heavy nucleus. The only way a proton can get close enough to fuse with a nucleus is if it is moving extremely fast, which requires an extremely high temperature. However, if the nucleus is very heavy indeed even the ultimate speed permitted by nature – the speed of light – is not enough to enable a proton to smash its way into the nuclear fortress.

The alternative to assembling atoms from protons is to assemble them from the other nuclear building block: the neutron. Because the neutron is without any electrical charge, it has the potential to slip easily through the walls of the nuclear fortress.

The first person to investigate the possibility that the elements had been made by such a neutron build-up process was von Weizsäcker in 1937. He proposed that somewhere out in the universe there was a place where atomic nuclei were relentlessly bombarded by neutrons.

A nucleus subjected to such a bombardment would swallow neutron after neutron until eventually it would become so neutron-rich that it became unstable and would beta decay, converting one of its neutrons into a proton. The resulting nucleus – now of an element one place higher in the Periodic Table – would in turn gobble neutron after neutron and the whole process would repeat itself.

Von Weizsäcker hoped to show that a long series of such neutron additions, punctuated by beta decays, could build up all the elements in nature. If he succeeded, it would be up to the astronomers to find a place in the universe with the necessary supply of free neutrons.* On the face of it, such a neutron build-up scheme was a nightmare to imagine. Once under way, there would be literally hundreds of isotopes of scores of elements, all in the process of swallowing neutrons or undergoing nuclear decay. However, the reality is even more complicated.

If a light nucleus gains a neutron, the resulting nucleus generally possesses less energy per nuclear building block.† In other words, the nucleus slides downhill towards the bottom of the curve of binding energy, shedding its surplus energy as it goes. If the heavier nucleus which forms were then to lose a neutron, restoring the original light

---

* This was a significant challenge, since a neutron left to its own devices self-destructs within about 10 minutes!

† The reverse is true, of course, if the nucleus is heavy and on the other side of the valley of stability. For such a nucleus, the opposite of the following explanation applies.

nucleus, it would have to go uphill away from the valley bottom and energy would have to be supplied from outside. It would seem reasonable to assume that this process never happens. However, this assumption would be wrong. This is because the nuclei involved in element-building would be flying about in a chaotic gas of particles. Each nucleus therefore carries energy of motion which can be transferred to another nucleus during a collision. This can drive the uphill process, reversing the results of any neutron build-up scheme.

Indeed, any nuclear reaction that can go in one direction can equally well go in the opposite direction. Consequently, a process in which neutrons are added to a nucleus to make a heavier nucleus will have to compete with the reverse process, in which neutrons are lost from the heavier nucleus. Once a scheme like von Weizsäcker's is under way, the hundreds of isotopes of scores of elements spit out neutrons as well as swallowing them. With the mix of elements changing from instant to instant, it seems impossible that such a process could ever create the constant mix of elements observed in the real universe.

However, change and constancy are not incompatible. They can, and often do, exist side by side. For instance, our bodies continually lose heat into the air but our body temperature stays roughly constant because heat is replaced as quickly as it is lost. Similarly, water is continually lost from the sea by evaporation, but the sea level stays constant because rain and rivers replace the water as quickly as it is lost.

Physicists call this kind of situation – in which things change constantly but stay the same – 'dynamic equilibrium'. In the case of von Weizsäcker's neutron build-up scheme, the mix of elements stays the same if, on average, every nucleus gains neutrons exactly as fast as it loses them. Remarkably, such a circumstance – thermodynamic equilibrium – is possible.

In thermodynamic equilibrium, despite all the frenzy of nuclear reactions going on, the overall mix of elements does not change with time: it 'freezes out'. And, crucially, the mix which freezes out is predictable.*

The essential prerequisite for establishing thermal equilibrium is that the particles of a gas interact with each other extremely frequently. This is because the state in which every nuclear reaction is balanced by its opposite is attained by a multitude of small energy adjustments, and such energy adjustments can occur only when particles collide. Since particles interact very frequently only when they are pressed close together, the usual prerequisite for thermal equilibrium is a high density.

---

* Strictly speaking, temperature has a meaning only for matter in thermodynamic equilibrium.

For the kind of nuclear thermal equilibrium envisaged by von Weizsäcker, it is also necessary that the temperature be extraordinarily high. This is because the nuclei have to be moving around with at least enough energy to drive the nuclear reactions one step uphill. The difference in binding energy between adjacent nuclei – ones differing by a single neutron – is very large. It corresponds to a temperature of billions, or even tens of billions, of degrees. If von Weizsäcker was correct and the neutron build-up process was responsible for the atoms in our bodies, somewhere in the universe there had to exist a magic furnace at least a hundred times hotter than the heart of the sun.

## THE FAILURE OF NUCLEAR THERMAL EQUILIBRIUM

In thermal equilibrium, the relative abundances of the different elements depend only on the temperature, density and binding energies of the nuclei, and can be calculated by inserting these numbers into a relatively simple formula. The abundances are entirely independent of the complex sequence of nuclear reactions that have brought about thermal equilibrium. For thermal equilibrium, history is bunk. This is its beauty.

Von Weizsäcker set out to find whether the observed abundances of the elements could have been produced in a furnace in nuclear thermal equilibrium. He began by considering the isotopes of single elements such as neon and silicon. He discovered that the observed abundances of neon's isotopes could have been created at a temperature of about 3 billion degrees. The corresponding temperature for silicon was nearly 13 billion degrees. It was not a good agreement. However, given the great uncertainties in the estimates of cosmic abundances, the temperatures could be regarded as being in the same general area, and compatible with the existence of a single furnace somewhere out in space.

However, problems arose when von Weizsäcker calculated the necessary densities. For silicon, for instance, the necessary neutron density was a million times greater than for sulphur. The figures were not even remotely comparable. The message in the element abundances was unequivocal. The mix of elements in the universe simply could not have been forged in thermal equilibrium – that is, in a single furnace with a unique density and temperature.

With the benefit of hindsight, it is clear that the very shape of the curve of binding energy makes it impossible for all the elements to have formed in thermal equilibrium. If the temperature of a gas of nuclei and neutrons was extremely high – more than, say, about 10 billion degrees

– collisions between fast-moving particles would tend to drive nuclei far uphill, away from the valley bottom, thus preventing light nuclei from ever acquiring neutrons and becoming heavier. If thermal equilibrium had been achieved at an extremely high temperature, therefore, the universe would contain only light atoms. On the other hand, if the temperature of the gas were relatively low – less than about 10 billion degrees – collisions in the gas would be incapable of stopping nuclei from sliding downhill to the valley bottom. So, if thermal equilibrium were achieved at a lower temperature, the universe would contain only middle-weight atoms.

Von Weizsäcker's failure to prove that the elements could have been made in thermal equilibrium nevertheless revealed an important truth about the creation of the elements. Each element must have been cooked at a different temperature and density. Whatever the cosmic furnace that forged all atoms, it therefore had to support a very great extreme range of temperatures and densities.

The obvious place to find such a furnace was deep in the heart of stars, since stellar cores are both hot and dense. However, the 10 million degrees or so found in the centre of the sun falls woefully short of the billions of degrees required for forging all the elements. Other stars were also ruled out since Eddington's theory indicated that they shared interiors broadly similar to the sun.

One possibility remained. If the pressure at the centre of a star were suddenly to vanish – something which could happen if the fuel supply ran out – gravity would cause the star to shrink out of control. As it dwindled down to a speck, getting denser and hotter by the second, the extreme conditions necessary to create each element might be created in sequence.

The problem was that if any newly forged elements were to provide the raw material for new suns, planets and the rock beneath our feet, they must somehow be thrust out into space. However, there was no known force that could prevent gravity crushing a star into the ultimate collapsed object – a black hole. The freshly forged elements would be locked away forever.

But if the magic furnace that forged all atoms was not to be found inside stars, where else was it to be found? One man had an extraordinary hunch. That man was George Gamow.

# The Inferno at the Dawn of Time

## WERE THE ELEMENTS MADE IN THE FIREBALL OF A HOT BIG BANG?

The elements were cooked in less time than it takes to cook a dish of duck and roast potatoes.

George Gamow

George Gamow, like Carl-Friedrich von Weizsäcker, was convinced that atoms had been made in a furnace somewhere out in space. When it became clear in the late 1930s that element-building required an extremely wide range of conditions – far wider than was available inside stars – Gamow began to look elsewhere for the magic furnace.

But where did the temperature routinely exceed a billion degrees? It was Gamow's great insight to realise that the entire universe had once been this hot – shortly after its birth.

The idea that the universe had been born at some time in the past, and had not existed forever, was a consequence of two sensational discoveries made in the 1920s by the American astronomer Edwin Hubble. Using the world's biggest telescope, on Mount Wilson in California, Hubble had found that the Milky Way – the great island of stars which includes the sun – was merely one 'galaxy' among countless billions of others suspended in the ocean of space. In 1929, Hubble followed up his discovery of galaxies – the fundamental building blocks of the universe – with the discovery that the universe was expanding, its galaxies flying apart from each other like pieces of cosmic shrapnel.

If the universe was expanding, one conclusion appeared inescapable: it must have been smaller in the past. There must have been a moment when all of creation occupied a minuscule volume of space: the moment of the universe's birth. By imagining the expansion running backwards, like a movie in reverse, it was possible to deduce that the universe had exploded from this super-dense state many billions of years ago.

The evidence for a 'big bang' seemed incontrovertible. However, when Hubble used his estimate of how quickly galaxies were receding from each other to deduce how long it had taken them to get to their present separations, he came up with a timespan of only 2 billion years. But this was less than the age of the earth, whose most ancient rocks had been dated at least 4 billion years old. Faced with the ridiculous prospect of a universe younger than the earth, hardly anyone was prepared to accept the idea that the universe had begun in a titanic explosion billions of years ago. Hardly anyone, that was, except George Gamow.

Gamow, now resident in the United States, was in a unique position to take the big bang seriously. At the University of Leningrad he had worked for Aleksandr Friedmann, one of the first people to use Einstein's theory of gravity to determine what kind of universe we lived in. Seven years before Hubble's discovery of the fleeing galaxies, Friedmann had concluded that we lived in a big bang universe.

But believing that the universe had been born in a big bang was one thing. Speculating on what the earliest moments of creation had actually been like was quite another. The newborn cosmos was so unimaginably remote from everyday experience as to be almost unthinkable. However, thinking the unthinkable was what Gamow did best.

The first thing Gamow realised was that the early universe must have been hot. If it were possible to run the expansion backwards in time, so that the universe was squeezed into a smaller and smaller volume, it would get hotter and hotter just as the air in a bicycle pump gets hotter and hotter the more it is compressed. Taking the argument to its logical conclusion, at the moment of the universe's birth, when all of creation was squeezed into an unimaginably small volume, the universe must have been unimaginably hot.

Then, as the universe expanded rapidly from its super-compressed state, it would have rapidly cooled. In fact, as it expanded and cooled, it would have spent a brief time at every conceivable density and temperature. It was Gamow's genius to realise that an extraordinary range of conditions would have been created in the earliest moments of the expanding universe – a far wider range than was available inside any star.

This was precisely what Gamow was looking for. He proposed that the atoms in our bodies had been forged in the searing-hot explosion of the big bang. The magic furnace had existed at the very dawn of time.

# A DISH OF DUCK AND ROAST POTATOES

Gamow assumed that the universe had been filled with some simple initial ingredient from which all atoms were subsequently cooked. There were only two possible candidates for the initial ingredient – the two nuclear building blocks, protons and neutrons.

A universe initially filled with protons was an impossibility, since the tremendous electrical repulsion the particles exerted on each other prevented them from sticking together to make heavy elements. Neutrons seemed more feasible, but some would have had to decay into protons before any element-building could begin, since all atomic nuclei except hydrogen contain protons. The problem was that once a significant number of neutrons had decayed into protons – something which typically took about 10 minutes – there was only a brief period in which the elements could be built up before all the neutrons had decayed.

The obvious solution was for the universe to start off with both protons and neutrons so that element-building could get going right away. This was what Gamow came to favour – an initial mix consisting of protons and neutrons in equal proportions. In the immediate aftermath of the hot big bang, he suggested, neutrons began to collide with and stick to protons. By a series of such neutron additions, punctuated by beta decays, heavier and heavier nuclei had then been built up as the universe expanded and cooled.

Gamow's scheme was still severely limited by the short lifetime of the free neutron.* It was a race against time. The initial ingredients of the universe had to be processed into the multitude of elements which abound in today's universe. And the frenzy of nuclear reactions that accomplished the feat had to be over and done with in little more than 10 minutes.

The 10-minute time-scale completely determined the conditions necessary for element-building in the big bang. For there to be appreciable element build-up, Gamow reasoned, a good fraction of neutrons would have to collide and stick to a proton before they decayed. In other words, there had to be a proton somewhere along the flightpath of each neutron in its 10-minute lifespan.

---

* Von Weizsäcker had required a constant neutron density for his element build-up process. He was therefore tacitly assuming that neutrons were being replenished by an 'unspecified' outside source as quickly as they self-destructed. No such possibility was open to Gamow, since his furnace was the entire universe and, by definition, there was no outside. Gamow was therefore confined to the neutrons that had existed at the beginning of the universe.

A neutron is like a little bullet which, as it flies, sweeps out a cylindrical volume of space corresponding to its cross-section. (Think of a bullet shot through a slab of butter: the volume of space swept out is simply the volume of the space left behind.) The volume of space swept out by a neutron in 10 minutes depends on its speed. This, in turn, depends on the temperature, which von Weizsäcker had determined must have been at least a billion degrees for neutron build-up to be viable. It was a relatively simple matter to estimate from this how many protons there had to be in every cubic centimetre, and therefore the density of matter at the time of element-building. Gamow found that it was a millionth the density of water.

But in addition to matter, the early universe contained radiation – light. How much of this there was again depended on the temperature. Gamow worked out the energy density of radiation at a temperature of a billion degrees and compared it with the energy density of matter.* He found that the energy in radiation was 10 million times greater than the energy in matter.

In the big bang, radiation was king. Light radiation controlled how quickly the universe was expanding and, consequently, how quickly the temperature was falling. Knowing this, Gamow was able to calculate precisely how long the exploding fireball spent at any particular density and temperature. This calculation provided an important insight into the nuclear process going on in the big bang.

Von Weizsäcker's calculations of nuclear thermal equilibrium had shown that for each element a unique combination of temperature and density would ensure the creation of its isotopes in the relative abundances seen in nature. However, he was tacitly assuming that there was always sufficient time for a mixture of nuclei and neutrons to settle down to the state of nuclear thermal equilibrium.

How much time constituted 'sufficient time' would depend on beta decay, since this is instrumental in bringing every nuclear process and its reverse into balance. Usually, nuclei waited minutes before they beta-decayed. However, Gamow estimated that in the rapidly expanding fireball of the big bang, the temperature and density needed to make the isotopes of a particular element existed typically for less than a second. There was simply no time to establish the state of nuclear thermodynamic equilibrium. The early universe was changing too rapidly.

Gamow's insight complicated matters. After all, in nuclear thermal equilibrium the abundances of the different elements depended only on

---

* The energy density of matter, using Einstein's famous relation, is simply the density of matter multiplied by the square of the speed of light.

the temperature, density and binding energies of the nuclei, and was straightforward to calculate. The abundances did not depend on the precise sequence of nuclear reactions which had led to the establishment of equilibrium.

By contrast, in the 'non-equilibrium' situation of the big bang, the sequence of reactions was of crucial importance. To discover the mix of elements created, there was no option but to follow through the complex chain of neutron additions and beta decays which had gone on up until the moment the last free neutron self-destructed.

As it turned out, the lifetime of the neutron was not the only factor limiting element-building. Radiation further constrained it. The high energy density of radiation in the big bang translated into a huge number of photons – the bullet-like particles of light. In fact, according to Gamow's calculations, there were a billion photons for every nuclear building block – an astonishing number. It was clear that protons and neutrons were no more than occasional flotsam in an overwhelming sea of photons.

Each photon, furthermore, carried a large amount of energy. In fact, when the temperature of the universe was above 10 billion degrees, the energy carried by each photon was sufficient to blast apart a newly formed atomic nucleus. Element-building was therefore impossible at that temperature. The process could start only when the temperature in the cooling fireball had dipped below 10 billion degrees, a critical threshold not passed until the universe was about 10 seconds old.

The task facing Gamow was now completely defined. He had to show that an orgy of element-building nuclear reactions – starting when the universe was about 10 seconds old and finishing when it was about 10 minutes old – could turn a mix of protons and neutrons into the multitude of elements we see around us today.

But Gamow had neither the patience nor the aptitude to carry out such calculations. He gave the problem instead to two young physicists, Ralph Alpher and Robert Herman.

## THE PREMATURE DEATH OF THE BIG BANG

Alpher had met Gamow at George Washington University, where Gamow was a professor on the faculty and Alpher was taking evening classes. Their paths next crossed at Johns Hopkins University in Baltimore, where Alpher obtained a post after leaving the navy in 1944. Gamow was appointed a consultant to Johns Hopkins the following year and Alpher promptly signed up as his PhD student.

In 1946 Gamow suggested that Alpher explore the idea that the elements had been made in the first few minutes of a hot big bang. In the next-door office was a young physicist called Robert Herman. His conversations with Alpher led to Herman becoming so fascinated with the idea of cooking up atoms in the big bang that he dropped what he was doing and joined Alpher. It was the start of a 50-year partnership.

Since the whole element build-up process depended on adding neutrons to atomic nuclei, one at a time, it was crucially important to know how enthusiastically different nuclei snapped up any neutrons they encountered. If a nucleus was too slow, a neutron might fly on past before it could react and make a heavier nucleus; if a nucleus was fast, on the other hand, neutrons might be gobbled up, enabling element-building to proceed at breakneck pace.

The propensity of a nucleus to swallow neutrons is technically known as its neutron-capture 'cross-section'. It had been measured for many nuclei as part of the programme to build the atomic bomb.* In 1945, everything surrounding the bomb project was a closely guarded secret. Fortunately for Alpher and Herman, however, the US government began declassifying its bomb data in 1946. However, there were some gaps in the information. The bomb scientists had not measured the neutron-capture cross-sections for all nuclei. Alpher and Herman were reduced to guessing the data they did not know from the data they had.

Calculating element build-up in an expanding universe was extremely difficult without electronic computers. Nevertheless, Alpher and Herman soon scored a dramatic success. They found that nuclear reactions in the furnace of the big bang converted roughly 25 per cent of the mass of the universe into the helium, the second lightest element. It was exactly the proportion the astronomers had observed throughout the universe. However, despite this success, Alpher and Herman registered a spectacular failure with all the elements heavier than helium. None, it seemed, could be made in the furnace at the beginning of time.

What stood in the way of the element build-up process was a missing nucleus. Nature had omitted to provide a stable nucleus made from 5 nuclear building blocks. It meant that once the neutron build-up process had made helium-4 there was nowhere for it to go. There was no prospect of making heavy elements.

As it turned out, nature had conspired to block more than the neutron build-up process. For an obvious way of leapfrogging the mass-5 barrier might have been for two helium-4 nuclei to collide and stick to

---

* This was done in order to determine the suitability of various materials for the construction of a nuclear reactor, in order to make the bomb fuel plutonium.

make a nucleus of mass 8. Unfortunately, nature had not only omitted to provide a nucleus made of 5 nuclear building blocks but also one made of 8!

The route to building heavy elements in the hot big bang was well and truly blocked. Despite Gamow's great hopes, the billion-degree fireball at the beginning of time could not have been the magic furnace. The pendulum had already swung from stars to the big bang. Just when it seemed there was nowhere else left to look for the elusive magic furnace, the pendulum swung back towards the stars.

# I I

# *The Key to the Stars*

HOW WE DISCOVERED THAT THE STARS
CONTAIN THE KEY TO UNLOCKING THE
SECRET OF ATOMS, AND THE ATOMS
THE KEY TO UNLOCKING THE
PUZZLE OF THE STARS.

I think the stars are the crucibles in which lighter atoms are compounded
into more complex elements.

Arthur Eddington

The dead end reached by Gamow was a major problem, since there were no other furnaces in the universe sufficiently hot and dense to forge the elements. Atoms simply had to have been made in either the big bang or the stars. In the light of this, it seemed a good time to take a fresh look at stellar interiors. How confident were the astronomers that they really knew what it was like deep inside stars?

Arthur Eddington had theorised that a star's total light output depended essentially on its central temperature, which in turn depended on its mass. This prediction was in perfect accord with observations of stars like the sun, whose central temperatures were calculated to be tens of millions of degrees. Although this was far too cool to forge heavy elements, it was nevertheless perfectly adequate for welding hydrogen nuclei together to make helium, a process that by liberating copious amounts of nuclear binding energy could account for sunlight.

A theory which had notched up so many remarkable successes was unlikely to be completely wrong. Nevertheless, Eddington's brainchild was not without its problems. Most worryingly, it was completely unable to account for the existence of red giants. This failure was especially ironic since it was red giants on which Eddington had first focused his attention, believing that their obvious gaseous nature made them potentially understandable. Subsequently, of course, it had turned out that even the sun was made of 'ionised' gas, or plasma, and that Eddington's theory in fact described most stars. Most stars, that was, except red giants.

Not only did Eddington's theory fail miserably to account for their size and prodigious light output, it actually implied that these stars should not exist!

## GIANT PROBLEMS

Red giants combined coolness with prodigious light output, properties which Henry Norris Russell had realised could be reconciled if the glowing surfaces of such stars were sufficiently vast. But it was the tremendous size that led to problems. It implied that a lot of the matter in such a star is spread a long way from its centre. And since gravity gets weaker and weaker at ever greater distances from a concentration of mass, much of the outer envelope of a red giant must weigh very little.

But the weight of the star bearing down on its central regions is what determines the temperature in the core. If the outer layers of the star weigh relatively little, it follows that the gas in the core need only have a relatively modest temperature in order to support those layers – only a few million degrees, compared with the 15 million degrees in the case of the sun. But here is the problem. A temperature of only a few million degrees is far too cool to drive the hydrogen fusion reactions that power stars. Red giants had no right to be shining at all.

In desperation George Gamow, a man renowned for his powers of lateral thinking, suggested that red giants were running on a nuclear fuel other than hydrogen. He suggested lithium, the third lightest element, since it was known to 'burn' at a lower temperature than hydrogen. Unfortunately, lithium was not common in stars, and Gamow was reluctantly obliged to abandon the whole idea.

Red giants remained an astronomical enigma. No one knew what powered them or what puffed them up to such enormous dimensions. It was clear that a new idea was needed to break the theoretical deadlock. Just such an idea was provided by two young theorists working in England – Ray Lyttleton and Fred Hoyle.

# A PEACH OF A STAR

Fred Hoyle was an unlikely academic. As a Yorkshire schoolboy he had frequently played truant and had taught himself to read in the back seat of his local cinema. Against all the odds, however, he won a scholarship to Cambridge in the late 1930s, graduated with the highest honours, and became the student of Paul Dirac, widely regarded as the greatest English physicist since Newton.

Dirac had not wanted a student and Hoyle had not wanted a supervisor. A member of the Cambridge faculty who knew about this had put them together as a joke. But although theirs was not a match made in heaven, Hoyle benefited greatly from his contact with Dirac. For the great man gave him a piece of advice, which he took to heart and which was to set the course of his scientific life.

Dirac informed Hoyle that the golden age of physics – the quantum revolution in which Dirac himself played a pivotal role – was at an end. If Hoyle wished to make his name by solving important problems, he should therefore look beyond physics to more fruitful fields of science. For Hoyle, it was a toss-up between biology and astronomy. He chose astronomy for the simple reason than it was first to throw up an interesting problem – or rather an interesting collaborator. His name was Ray Lyttleton and he was a young Cambridge astronomer. Hoyle bumped into Lyttleton in 1939 and the pair hit it off right away. They teamed up and began working together on the theory of stars.

Thanks to Hans Bethe in the US and Carl-Friedrich von Weizsäcker in Germany, the two English theorists had an advantage over Eddington. They knew the precise nuclear fusion reactions that powered stars, and how the heat these generated depended on the temperature and density in a stellar core. By incorporating this new information into the theory of stars, they were able to go an important step beyond Eddington.

To predict the total light output of a star, Eddington had needed to know two things – the star's mass and surface temperature. Hoyle and Lyttleton were able to dispense with the surface temperature. From a star's mass alone, they were able to predict both its light output and its surface temperature. Furthermore, if a star's light output and temperature were known, it was always possible to deduce how big it was. After all, a star which was dim despite being hot had to be small whereas a star which was bright despite being cool had to be large. Hoyle and Lyttleton therefore had a powerful tool for determining the size of any star they might care to imagine.

They immediately applied it to the problem of red giants. Was there any way to concoct a star that would swell up to such monstrous dimensions? To their delight, Hoyle and Lyttleton found that there was. A star would inflate like a great red balloon if there existed a difference in composition between its core and outer regions – more precisely, if the core was made of heavier particles than the rest of the star.

Here then was the recipe for a red giant. It was a star whose interior was partitioned into inner and outer zones, each with a different composition, not unlike a peach with a dense stone in the middle. All that remained was to determine how a star might acquire its peach-like interior.

The obvious method is simply to grow old. After all, as a star ages, it transforms more and more of the hydrogen fuel in its core into helium. Since helium is four times as heavy as hydrogen, this boosts the average weight of the gas particles at the centre. It was possible to imagine the star becoming more and more peach-like with every passing year until, eventually, it swelled into a red giant. It was a case of a relatively simple change in a star's internal make-up leading to an absolutely spectacular change in its external appearance.

There was only one problem. According to the received astronomical wisdom, it was impossible for the heart of a star ever to become weighed down with helium. As Eddington had discovered, inside every star is a giant egg-beater – the consequence of stellar rotation. By blending together the ingredients of a star, it ensures that at all times its interior is like a well-mixed cake.

## THE EGG-BEATER INSIDE STARS

Most stars, including the sun, are spinning like tops. Their surfaces are therefore flung outwards like a child on a fast merry-go-round. The effect is most marked at a star's equator, where the material is whirling round most quickly. It gives stars a slight midriff bulge.[*]

This bulge affects the way in which heat flows in the star. Since the equator is slightly farther from the fires in the core than the poles, it is slightly cooler than the poles. Consequently, heat, which always flows from a hot place to a cold place, flows from the star's poles to its equator.

It might be expected that the heat would flow through the gas of the star, in much the same way that heat flows through the material of a metal bar. However, as Eddington discovered when he looked inside a spinning star in detail, maintaining the temperature balance within a star is actually a little more complicated. Instead of heat flowing through the

---

[*] The sun, for instance, is about 70 kilometres fatter at the equator than it is pole to pole.

gas, the gas physically moves from the star's hot regions to its cool regions, carrying the heat along with it.

Obviously, the flow cannot be one-way. This would cause gas to pile up in one part of a star and disappear completely from another, which is plainly ridiculous. The flow must instead be two-way, with the gas eventually returning to its starting point. Eddington therefore concluded that immense currents of gas must circulate endlessly in the bowels of stars moving quickly enough to mix the gas thoroughly in even a slow-spinning star. In short, they act like a giant egg-beater. The egg-beater takes freshly made helium from the core and stirs it throughout the bulk of a star. In this way, it prevents a peach-like interior from ever developing.

The egg-beater inside stars seemed to have dealt Hoyle and Lyttleton's ideas a fatal blow. But the two astronomers refused to be discouraged. They had hit on a promising solution to the red giant puzzle; there had to be a way round this obstacle. All they had to do was find it.

## TUNNELLING THROUGH SPACE

One possibility the pair considered required that the starlanes were littered with drifting clouds of hydrogen gas. Whenever a star happened to plough through such a cloud, it would inevitably sweep up, or 'accrete', some of the gas, which would accumulate on its surface.[*] If the cloud were dense enough, by the time the star emerged on the far side it would have gathered about it a thick coating of hydrogen gas. With an exterior of pure hydrogen, and an interior which was a mixture of hydrogen and the heavy helium ash of nuclear burning, the star would now have precisely the peach-like structure required to turn it into a red giant.

If Hoyle and Lyttleton were right, a star would spend only a limited time as a stellar extrovert before the egg-beater inside once more mixed together the gas in its interior. The star would then deflate back down to its original size and resume its former existence – until such time as it ran into another hydrogen cloud. The possibility of repeat encounters with such clouds meant that a star might become a red giant more than once in its life.

Hoyle and Lyttleton's scheme for making red giants was ingenious.

---

[*] The possibility that stars flying through space might sweep up enough gas and dust to modify their composition was the problem Lyttleton had been working on when he first met Hoyle, and which had piqued Hoyle's interest in astronomy.

However, it turned out to be completely unnecessary. By the time the two astronomers had worked out the details, an embarrassed Eddington had owned up to an uncharacteristic error in his calculations. He had grossly overestimated the speed of gas currents circulating within stars. In reality, they were extremely sluggish. They crept round the sun, for instance, at a mere hundred billionth of a metre per second – so slowly that it would take a whole year for them to travel the thickness of a strand of cotton, a human lifetime to cross the face of a wristwatch. The interior of a star could not conceivably be kept uniformly mixed by such ponderous churning.

Nothing at all therefore prevented helium ash from nuclear burning piling up in a stellar core. As it grew old a star would inevitably develop a peach-like interior. Even our own sun, five billion years from now, will eventually become a red giant.

In fact, Hoyle and Lyttleton's accretion mechanism would never have worked. It relied on the existence of cold clouds of hydrogen gas drifting between the stars. Although such clouds were discovered in the 1950s, they proved to be far too rarefied to coat a passing star with the necessary hydrogen to make it a red giant.* But the accretion process, although a red herring as far as red giants were concerned, turned out to be of central importance in astronomy. Some of the most powerful objects in the universe, such as X-ray binary stars and quasars, are now believed to be powered by black holes which are accreting gas from interstellar space.

## THE BIRTH OF A RED GIANT

With the red giant puzzle solved at last, a new picture emerged of the evolution of a star like the sun. It was infinitely richer than the one painted by Eddington.

The egg-beater, if it had existed, would have gradually diluted a star's hydrogen fuel with helium ash. The fires in the core would have dwindled until, eventually, the temperature was too low to drive energy-generating nuclear reactions. The fate of a star like the sun would therefore be to fade away and die. Such a picture of the evolution of stars like the sun could not be farther from the truth. Far from going out with a whimper, they set the heavens ablaze!

The detailed sequence of events that overtake a star as it grows old is

---

* Too cold and dark to be seen with conventional optical telescopes, the hydrogen clouds were detected by telescopes sensitive to radio waves, a type of light invisible to the human eye.

complicated. However, its chief characteristics would be discovered in the mid-1950s by Hoyle and the Princeton theorist Martin Schwarzschild.

Broadly speaking, the first crucial event occurs when the hydrogen fuel in the star's core becomes completely exhausted. The star is deprived of the energy-generating nuclear reactions which have sustained it over billions of years of normal life, and its centre is now occupied by a compacted ball of helium ash. Unable to replenish the heat it is losing, the ball cannot support its own weight against the pull of gravity. It begins to shrink.

The shrinkage heats up not only the core but also the layers of the star immediately above it. Since these layers contain unburnt hydrogen, they catch fire.* Hydrogen now blazes in a flaming ring, or 'shell', around the star's helium core. The core continues to contract and heat up, raising the temperature of both itself and its surroundings. Hydrogen reaches ignition point farther and farther from the centre, and the ring of fire moves outward through the star.

But the dramatic events which will transform the star into a swollen red giant are yet to occur. They are triggered by the huge quantities of helium ash that rain down on the core from the furiously burning hydrogen shell above. The helium ash increases the mass and gravity of the core, which in turn causes the core to shrink and heat up faster.

The increased gravity of the core has another effect. It claws down material from the hydrogen-burning shell, which is now the only thing supporting the crushing weight of the overlying layers of the star. It's as if someone who is trying to hold up a collapsing ceiling has weights tied to their straining arms. Somehow the shell of flaming hydrogen must increase its pressure in order to counterbalance the gravity trying to drag it down.

There are two options: either the hydrogen-burning shell can increase its temperature, or it can increase its density. In fact it does both. Either change, in isolation, would fan the flames of burning hydrogen; coming together as they do, they cause the shell to flare up with unprecedented fury.

The photons, which surge outwards through the star, are forced to follow the most tortuous of zigzag paths to the surface. They are dammed up inside the star. The blazing hydrogen shell is now generating energy far faster than the photons can transport it to the

---

* Herein lies the crucial difference between uniform stars, as envisaged by Eddington, and the non-uniform stars of Hoyle and Lyttleton. Immediately outside the core of a uniform star would be exactly the same helium-poisoned mixture as inside the core. There would be no virgin hydrogen waiting to ignite.

surface. The excess must go somewhere. In fact, it goes into heating the bulk of the star.

The star inflates. And, as the surface layers expand, they cool and turn red in colour. At this stage, the inert helium core contains about half the mass of the entire star. However, it is squeezed into only about one ten billionth of its volume. So a red giant is in fact two stars – a great globe of rarefied gas with a super-dense, super-hot helium ball of helium in its heart. It is a truth so bizarre that nobody, not even Eddington, would have guessed it.

## BUILDING ELEMENTS INSIDE STARS

The success of Hoyle and Lyttleton in explaining the most striking properties of red giants shattered forever the idea that a star's chemical ingredients stayed uniformly blended together at all times in its life. A star might be born as a globe of thoroughly mixed gas but, inevitably, as the helium ash of nuclear burning built up in its core it would become non-uniform in composition.

The development of non-uniformity with age was the key to understanding stars.

With increasing non-uniformity, the conditions inside a star become ever more extreme. In the super-dense helium core of a red giant the temperature might rise to 100 million degrees – ten times the temperature in the middle of the sun, and far higher than anything imagined by Eddington.

The implications for element-building were profound. In the blisteringly hot core of a red giant, it was possible to imagine helium nuclei slamming into each other so violently that they stuck despite the fact that they repelled each other far more strongly than hydrogen nuclei. The fusion of helium was a new source of nuclear energy for the star. It also resulted in nuclei of even heavier elements, such as carbon.

And there was no reason why element-building should stop at this point. For the carbon ash of helium burning would gravitate to the centre of the helium core for exactly the same reason that the ash of hydrogen burning had gravitated to the centre of the hydrogen core. The result would be a star with a core of carbon inside a shell of burning helium inside a shell of burning hydrogen. Instead of being merely divided into two zones with different compositions, the star would be divided into three.

And, once more, increasing non-uniformity would lead to increasingly extreme conditions within the star. For when the carbon ash had choked out the fires in the heart of the star and created a central ball of carbon, that too would shrink and heat up. In the blisteringly hot carbon core it was possible to imagine carbon nuclei slamming into each other so violently that they fused to make nuclei of an even heavier element, say magnesium.

The sequence of events as a star became ever more non–uniform was now clear. The exhaustion of nuclear fuel would lead to the shrinkage and heating of the stellar core, which would lead to the ignition of a new nuclear fuel. The exhaustion of that fuel in turn would lead to the shrinkage and heating of the stellar core, and so it went on. How many times this cycle would repeat depended only on the mass of the star, since it was the weight of the material bearing down on the centre of the star which ultimately determined how hot its core might get. In a very massive star, it was possible to imagine the cycle repeating over and over again, with the central temperature spiralling ever higher. The star would end up with a core of iron surrounded by shells of fusing silicon, oxygen, carbon, helium and hydrogen.* Rather than having an interior like a peach, it would have an interior like an onion!

If Hoyle and Lyttleton were right, element-building was the inevitable consequence of the evolution of stars. In fact, the evolution of stars actually drove the evolution of atoms, while the evolution of atoms in turn drove the evolution of stars. Once more, the stars were beginning to look like the 'magic furnace' where all the elements were forged. However, the forging of iron required a staggering temperature of 10 billion degrees. This was a full thousand times hotter than the heart of the sun. For most astronomers this was simply too ridiculous to contemplate.

What was needed was irrefutable proof that stars really could provide the staggeringly high temperatures needed to build the elements. Such proof would be found by Fred Hoyle in the penultimate year of the Second World War.

---

* Once the core of a star is made of iron the star has effectively run out of nuclear fuel since iron is the most tightly bound nucleus, and no more binding energy can be liberated by fusing it with other nuclei.

# The Ultimate Nuclear Weapon

## HOW WE FOUND DRAMATIC PROOF THAT THE EXTRAORDINARY TEMPERATURES AND DENSITIES NEEDED TO MAKE HEAVY ELEMENTS REALLY DID EXIST INSIDE STARS.

Now I am become Death, the destroyer of worlds.

From the *Bhagavad Gita*, quoted by Robert Oppenheimer
after the explosion of the first atomic bomb

Fred Hoyle's proof came about as a direct result of a wartime trip he made to the United States. At the time, he was working on the development of radar for the British navy at a secret research establishment near Nutbourne in West Sussex. It was in connection with this work that he was chosen to attend a radar conference scheduled to be held in Washington DC towards the end of 1944.

Hoyle crossed the Atlantic with 10,000 American troops returning home for Christmas. From New York, he caught the train south to Washington DC, where he found that the British embassy had drawn up a programme of visits for him. However, the programme did not start for three days. He could hardly believe his good fortune. He went back to the railway station and caught the first available train to Princeton, New Jersey.

What drew him north was the magnet of Princeton University. His war work had given him little opportunity to talk to anyone about astronomy. But at Princeton he might chat with one of America's greatest astronomers – Henry Norris Russell. Hoyle's meeting with Russell was every bit as stimulating as he had hoped. Over lunch Hoyle happened to mention that he was flying west to San Diego when the radar conference in Washington was over. If that was the case, the American astronomer replied, then he simply must visit the Mount Wilson Observatory near Los Angeles. He would write to the observatory's director right away and see to it that Hoyle was invited.

The radar conference came and went and Hoyle flew off to San Diego. In a spare weekend in his itinerary, he took Russell's advice and headed north to Los Angeles.

## BRIGHTER THAN A BILLION SUNS

The offices of Mount Wilson Observatory were in Pasadena, a city set among picturesque orange groves just to the northeast of Los Angeles. Hoyle arrived as a car was leaving for Mount Wilson. The observatory's director told him that if he wanted to jump in he could spend the weekend up on the summit. It was an offer too good to refuse.

Mount Wilson is a 5700-foot-high peak in the San Gabriel mountains, which towers above Pasadena and the Los Angeles basin. At its summit was the biggest telescope in the world – the 100-inch Hooker telescope. It was with the Hooker, in fact, that Edwin Hubble had created the modern picture of the universe, by identifying galaxies and discovering the big bang.

At the end of his weekend on the summit of Mount Wilson, Hoyle was met by a car waiting to take him back to Pasadena. In the driver's seat was the German astronomer Walter Baade. Baade was one of the greatest observational astronomers of the twentieth century. At the outbreak of war he had been classified as an enemy alien and excluded from military service. He had spent the war years with the Hooker telescope, taking advantage of the blacked-out skies above Los Angeles to probe the depths of the universe.

As Baade drove Hoyle down the wide palm-tree-lined boulevard to Pasadena, the two men talked of the latest developments in astronomy. Their conversation, which continued throughout the afternoon in Baade's office, touched on many subjects. Among them was the topic of 'novae', stars which underwent sudden and dramatic flare-ups. Far more interesting, Baade told Hoyle, were stars dubbed 'supernovae', which underwent flare-ups of such violence as to beggar belief. By comparison, novae were little more than penny candles.

Baade had been the first to recognise the existence of supernovae. In fact, it was Baade and his Swiss-American colleague Fritz Zwicky who had actually coined the term back in 1934. What had led the pair to suspect the existence of a distinct class of exceptionally brilliant novae was the fact that occasionally stars were observed to flare-up within 'spiral nebulae'. These misty patches of light littering the sky were generally assumed to be clouds of luminescent gas floating in space. Since glowing gas is too faint to show up at great distances, spiral nebulae had to be relatively nearby objects, which was entirely compatible with the stars erupting inside them being standard novae.

In the 1920s, however, Edwin Hubble pointed the 100-inch telescope on Mount Wilson at the constellation of Andromeda and resolved the largest spiral nebula into individual stars. The Great Nebula in Andromeda was not a cloud of interstellar gas. It was a galaxy – in fact, all the spiral nebulae were galaxies, far beyond the limits of our Milky Way. The only reason they looked like clouds of gas was that sheer distance had blurred their stars together.

The novae that flared up inside these galaxies could not therefore be ordinary novae. To be visible across such immense tracts of space they had to be extraordinarily bright. Baade and Zwicky had discovered that a supernova is a million times as powerful as an ordinary nova. In a brief moment of cosmic insanity, a single star flares up brighter than a hundred billion ordinary suns. It is a stellar conflagration of unimaginable proportions.

Hoyle asked for more details of supernovae, and Baade gave him copies of a number of recent scientific papers he had written on the subject, none of which had been available in wartime Britain. They would turn out to be the first ingredient in Hoyle's proof that stars really were able to forge the elements.

## THE PROBLEM WITH PLUTONIUM

The journey home took Hoyle to Montreal, from where it was now possible to fly all the way to Scotland in a giant Fortress or Liberator bomber. The weather in Montreal was too bad to fly, and Hoyle was forced to wait several days for it to clear. It was during this enforced stay that there occurred the second significant event of his transatlantic trip. Hoyle, although thousands of miles from home, bumped into two friends from Cambridge.

The first was Maurice Pryce, a high-flying theorist who had been an undergraduate with Hoyle. For a while, Pryce had been stationed at the Admiralty Signals Establishment, where Hoyle was working on radar. However, he had suddenly and mysteriously disappeared, and Hoyle had never been able to find out where he had gone. His second

Cambridge friend was Nick Kemmer, a nuclear physicist originally from Russia. Kemmer was part of Tube Alloys, a shadowy organisation which had been systematically recruiting university graduates with nuclear knowledge. Hoyle was in no doubt that Tube Alloys was the cover for the British project to build an atomic bomb.

The idea of building such a device had originated in 1939, when scientists in Germany had announced that a nucleus of uranium-235 splits into two when struck by a neutron. Since this 'fission' process was accompanied by the ejection of more neutrons, which in turn were capable of splitting more nuclei of uranium-235, it was suddenly possible to imagine triggering a runaway 'chain reaction' of fissions, which would unleash an unstoppable flood of nuclear binding energy. All that would be needed was a large enough lump of uranium-235.

Hoyle had assumed that the goal of Tube Alloys was to accumulate such a lump. The task was a formidable one because most natural uranium was uranium-238, not uranium-235, and the two isotopes were so alike that they could not be separated by any conceivable chemical means. The only obvious alternative open to Tube Alloys was to exploit the fact that uranium-235, being lighter than uranium-238, moved more swiftly. So, if a gas containing uranium was passed repeatedly through a porous material, what emerged would become richer and richer in uranium-235.*

The enrichment process was agonisingly slow. Accumulating enough uranium-235 for a bomb would probably take years. Could it be possible that Tube Alloys had succeeded in the task already? Hoyle considered it extremely unlikely. On the other hand, Tube Alloys had evidently transferred key personnel like Kemmer and Pryce to North America, where the new weapon would presumably be tested away from the prying eyes of the enemy. Everything suggested that the bomb project was nearing success.

It didn't make sense. Unless, it occurred to Hoyle, there was another, faster route to the bomb. If an uncontrolled chain reaction could lead to the explosive release of nuclear energy, a controlled chain reaction could lead to a more sedate release of nuclear energy. Such a controlled chain reaction would require the building of a device called a nuclear 'pile', or 'reactor'.

Inside such a reactor, free neutrons would collide with nuclei to make elements hitherto unknown in nature. For instance, uranium-238 would

---

* In fact, this 'gas diffusion' technique was precisely the one pioneered by Francis Aston before the First World War. It was his discovery that neon became lighter and lighter after repeated passages through porous clay that confirmed his suspicions that neon consisted of two main isotopes — a light isotope and a heavy one.

swallow a neutron and decay into neptunium. The discovery of neptunium had been announced in the June 1940 issue of *Physical Review*, the very last edition of the journal before all bomb-related research was classified. Neptunium was unstable, a fact realised in both Germany and America. It decayed into yet another man-made element: plutonium. Plutonium was the perfect bomb material. Not only did it undergo fission like uranium-235 but, since it was an entirely distinct element, it could be separated from uranium by relatively simple chemical means.

The snag was that the creation of plutonium required first building a nuclear reactor. Assembling such a device and simultaneously running a programme to concentrate uranium-235 was quite beyond the capabilities of war-ravaged Britain in 1940. It had been forced to choose one route to the bomb. As Hoyle twiddled his thumbs in 1944, waiting for the weather to clear above Montreal, it suddenly struck him how odd it was that Britain had chosen the most difficult route of the two. Why had it made work for itself? Only one answer made sense. Somewhere along the plutonium route to an atomic bomb there must be an obstacle.

With this thought, scattered jigsaw pieces began to come together in Hoyle's mind. He had heard a rumour that somewhere in the southwestern United States the Americans had assembled a bomb team from some of the greatest scientific minds in the free world.[*] It seemed odd to him that such a formidable team was needed simply to make a nuclear bomb explode. He had thought that triggering an explosive release of nuclear energy merely required taking two lumps of fissionable material whose combined mass exceeded the 'critical mass' for a runaway chain reaction, and slamming them together. The existence of the powerful American team was an indication that for plutonium this straightforward procedure did not suffice. For some reason, triggering a nuclear explosion in the man-made element must be more difficult than triggering one in uranium-235.

The only thing that could stop plutonium from exploding was plutonium itself. Perhaps, when two sub-critical pieces were brought together, the heat they generated was so intense that it drove them apart again before a full-blown chain reaction could catch hold. If so, the bomb-makers would have to force the plutonium to stay above the critical mass for sufficiently long to trigger a nuclear explosion.

The only sure way to do this, Hoyle realised, was to fashion the plutonium into a spherical shell and implode it in on itself. This could be

---

[*] The location of all the brain power, one of the best-kept secrets of the war, was Los Alamos in New Mexico.

done by surrounding the plutonium with a ring of conventional explosives; however, it was tremendously difficult. The explosives would have to be detonated absolutely simultaneously so that the implosion was perfectly uniform. The more he thought about it, the more Hoyle became convinced this must be the hidden obstacle on the plutonium route to an atomic bomb.

The idea that implosion was essential in order to trigger a nuclear conflagration would turn out to be the second crucial ingredient in Hoyle's proof that stars could easily be hot enough to forge all atoms.

## THE ULTIMATE NUCLEAR WEAPON

When the weather above Montreal finally cleared, Hoyle's plane took off for Scotland. From Glasgow, Hoyle took the train south to Nutbourne, arriving just in time for Christmas. It was only in the relatively quiet period between Christmas and New Year that he finally had the time to sit and reflect on the kaleidoscopic events of his trip to the US. Rereading the papers Baade had given him, he remembered speculating that implosion was essential to trigger the explosion of an atomic bomb. Could it be, he wondered, that implosion was also essential to trigger the explosion of a supernova?

At first sight it seemed a bizarre connection to make. After all, there was an enormous difference in scale between the explosion of a bomb and the explosion of a star. How could the two phenomena possibly have anything in common? However, if the implosion of a star were behind a supernova, it would certainly explain one thing: where the energy of the explosion came from. The shrinkage of a star would inevitably be accompanied by the release of gravitational energy, the fact which in the nineteenth century had led scientists to speculate that the sun maintained its temperature by slowly contracting. If a star were to implode, shrinking in size dramatically, the amount of gravitational energy unleashed would be truly tremendous. It would be easily enough to drive the explosion of a supernova.[*]

Just how the implosion of a star might come about was not too difficult to imagine. A very massive star, exhausting one nuclear fuel after another, and building up heavier and heavier elements in the

---

[*] In guessing that the implosion of a stellar core powered a supernova, Hoyle was not alone. Baade and Zwicky had come to the same conclusion in a classic paper on supernovae back in 1934. There was of course the small matter of how implosion was turned into explosion. But this was a detail which Hoyle believed would be sorted out in due course. Half a century later, supernova theorists are still trying.

process, would eventually end up with a core of iron. Since iron lies at the very bottom of the valley of nuclear stability, no more binding energy could be liberated by turning it into still heavier nuclei. On the contrary, such reactions would actually suck heat out of the core, causing a catastrophic drop in central pressure. With nothing to oppose the crushing force of gravity, the star would implode.

It was what would happen inside such an imploding star that fascinated Hoyle. As the material in the core was crushed into a smaller and smaller volume, its temperature would soar higher and higher. Matter would very quickly approach densities wildly in excess of anything found inside a normal star, while the temperature rocketed to billions upon billions of degrees. These were conditions the like of which nobody had dreamt might exist in stellar interiors. Once Hoyle realised they were a natural consequence of implosion, it changed his ideas about stars forever. Stellar interiors, long dismissed as too cool for significant element-building, might after all be the sites of the elusive magic furnace.

Everything of course hinged on the idea that a massive star, arriving at the end of its life, would implode. But though in principle such a catastrophic shrinkage could provide the energy to power a supernova, this was no guarantee that it did in practice. If Hoyle were to make a convincing case that stars were the site of the magic furnace, he would need to find direct proof that the staggeringly high densities and temperatures needed to cook atoms really were created inside supernovae.

## THE ASH OF LONG-DEAD STARS

The sort of super-hot conditions Hoyle envisaged for an imploding stellar core would cause a furious storm of nuclear reactions to rip through the star. In the ensuing chaos, atomic nuclei which had been painstakingly assembled over an entire stellar lifetime would slam into each other violently and either break apart into lighter nuclei, or stick together to make heavier nuclei.

The outcome of such a storm of nuclear reactions appeared impossible to predict. However, appearances could be deceptive. There was one very special circumstance, Hoyle realised, in which the outcome of nuclear reactions, no matter how complex, was always predictable. That very special circumstance was nuclear thermal equilibrium.

In nuclear thermal equilibrium, nuclear processes make each species of nucleus just as fast as they unmake it. With the forces of creation and destruction so perfectly balanced, the result is a 'steady state' in which the quantities of every type of nucleus become 'frozen' and unvarying with time, despite the chaos reigning all about. The key prerequisite for establishing nuclear thermal equilibrium is that the nuclear reactions are extremely fast and violent. It is a condition satisfied only at ultra-high temperatures, when nuclei are flying about extremely rapidly, and at ultra-high densities, when the distance that nuclei travel between collisions is short.

These are precisely the kind of super-hot, super-dense conditions which Hoyle envisaged for the interior of an imploding star. So if implosion drives the explosion of a star, it follows that nuclear thermal equilibrium exists inside supernovae. Furthermore, the relative abundances of the different nuclei in nuclear thermal equilibrium are entirely predictable. They depend only on the binding energies of the nuclei, together with the temperature and the density. So the task of determining whether nuclear thermal equilibrium really exists inside supernovae simply amounted to asking whether the abundances of nuclei observed in the universe could be produced by nuclear thermal equilibrium at a particular temperature and density.

Answering such a question required a knowledge of both the abundances of different elements and the binding energies of different nuclei. However, Hoyle had no access to such data, marooned as he was in West Sussex. Fortunately, however, war-related work took him to Cambridge in March 1945.

## THE 5-BILLION-DEGREE SIGNATURE

Once in the university town, Hoyle headed straight for the Cavendish Laboratory. In the building where J. J. Thomson had discovered the electron and Ernest Rutherford had carried out his pioneering experiments on radioactivity, he was confident that he would find a table of nuclear masses from which he might calculate the all-important nuclear binding energies.

Straight away, Hoyle ran into Otto Frisch, the Austrian physicist who in 1939 had first alerted the British government to the danger that the Germans might build an atomic bomb. Frisch had recently returned from Los Alamos and as luck would have it had exactly what Hoyle wanted. From the drawer of his desk, he pulled out a table of nuclear masses which had been compiled by a German nuclear physicist called Josef Mattauch.

Back in West Sussex, Hoyle set about calculating the elemental abundances that would be created in nuclear thermal equilibrium inside a supernova. He already had some idea of the result he would obtain. In the super-hot, super-dense conditions of nuclear hell, the nuclei whose production would be favoured would be those that were most tightly bound. Since the most tightly bound nucleus of all was iron-56, Hoyle guessed that the nuclear reactions inside a supernova would rapidly transform almost everything into iron, and other elements with a roughly similar atomic mass. The transformation into so-called iron-group elements was as inevitable as a ball rolling downhill to the most stable position at the foot of a valley. In fact, the analogy was an excellent one. The nuclei would gravitate to the bottom of the valley of nuclear stability.

Hoyle's hunch was borne out. In nuclear thermal equilibrium, the elements which were by far the most abundant ranged from scandium, with a mass about 45 times that of hydrogen, through iron-56, to nickel with an atomic mass of about 60. These were precisely the elements which had made human civilisation possible, which surely could be no coincidence. Hoyle then plotted the abundances of these elements against their masses. What emerged was a mountain peak. It rose from gently sloping foothills in the vicinity of scandium, up through titanium, vanadium, chromium and magnesium to a sharp peak at iron-56, the most abundant of the iron-group elements. On the far side of the peak, the mountain plunged steeply down through cobalt and nickel to foothills in the neighbourhood of copper and zinc.

Hoyle compared his findings to a table he had found in a book which he had borrowed from the main library at Cambridge. It was written by the Swiss-Norwegian chemist Victor Goldschmidt. In 1937, Goldschmidt had carried out a pioneering study of the composition of the universe. Pulling together data from the earth's crust, the sun and meteorites, he had compiled a table showing which elements were common and which were rare. In fact, the table had been used by von Weizsäcker in 1938. It was the German scientist's failure to reproduce Goldschmidt's relative abundances that had convinced him that nuclear thermal equilibrium could not have been responsible for creating all the elements in nature.

Goldschmidt had tabulated the relative abundances of elements from sodium, with an atomic mass of 23, upwards. When Hoyle plotted the abundances against the masses on a second piece of graph paper, what emerged was a slope which fell away steeply as elements got heavier and heavier. A heavy element like silver, for instance, with an atomic mass of about 108, was a million times rarer than a lighter element like magnesium, with an atomic mass of about 24. However, there was a notable exception to the general decrease in abundance with increasing nuclear mass. Between elements with an atomic mass of about 45 and 60 there rose a sharp peak.

Hoyle compared his graphs. The two peaks were absolutely identical. Here was unequivocal proof that the iron-group elements had been forged in nuclear thermal equilibrium.

The temperature Hoyle had used in his calculations was 5 billion degrees. It meant that somewhere out in space – probably in the hearts of supernovae – there really was a blistering furnace a thousand times hotter than the centre of the sun.

In 1946, Hoyle published a paper in which he proposed that, after a massive star had used up its nuclear fuel, the core would collapse, heat up and forge heavy elements, including iron. If Hoyle was right about supernovae, then the implications for stellar interiors were profound. Stars were known to be able to generate a solar-type temperature of 15 million degrees. If they were also able to generate a temperature of 5 billion degrees – admittedly only in the rare instances when they happened to blow themselves to pieces – then surely it was probable that they could generate all the intermediate temperatures between 15 million and 5 billion degrees? If so, then they were easily capable of forging all the elements in nature.

Hoyle had not the slightest idea of the complex sequence of nuclear reactions that stars might employ to build up heavier and heavier elements. However, if the sequence could be discovered it might be possible to explain not only the 'iron peak' but all the other features in Goldschmidt's abundance data. In short, it might be possible to explain the origin and abundances of every element.

## THE ORIGIN OF THE ELEMENTS

Of course, everything that Hoyle had done could have been done by Carl von Weizsäcker in 1938. However, von Weizsäcker had had no answer for the astronomers when they told him that temperatures of billions of degrees could not be found in stars. In addition, he was convinced that a super-hot witches' brew somewhere in the universe had spawned not just some but all of the elements in nature.

Hoyle's genius was not only to identify the one place in the universe where nuclear thermal equilibrium might exist, but also to realise that nature was not simple, and that the forging of atoms in nuclear thermal equilibrium might be only part of the story of the origin of the elements.

It might explain the iron–group elements, but something radically different was needed to account for the rest of the elements in nature.

It was immediately clear that the 'something different' was going to be complicated. After all, nuclear thermal equilibrium was the only situation in which the abundances of the elements depended only on the temperature and the density. In all other possible circumstances, the abundances would depend on the precise sequence of nuclear reactions that had gone on before.

Hoyle's task was therefore to discover the sequence of nuclear reactions needed to take place in stars for the assembly of all the elements from helium up to those of the iron-peak. It was a daunting task. But it was by this complex sequence of 'non–equilibrium processes' that most of the elements in our bodies, such as carbon, magnesium and calcium, had been forged.

The way ahead was clear. Bethe and von Weizsäcker had discovered the nuclear reactions that turned hydrogen into helium inside stars. The next problem was to find the nuclear reactions that could turn helium into heavier elements still. Unfortunately, there was a big obstacle blocking the way. That obstacle was the element beryllium.

# 13

# *Beyond the Beryllium Barrier*

## HOW THE MAJOR OBSTACLE TO BUILDING ELEMENTS IN STARS WAS FINALLY REMOVED.

As we look out into the universe and identify the many accidents of physics and astronomy that have worked together to our benefit, it almost seems as if the universe must have known that we were coming.

Freeman Dyson

George Gamow's heroic attempt to show that all the atoms in nature could have been made in the first minutes of the universe's existence had ended in failure. Although the fireball of the big bang could easily have cooked large quantities of helium, there was no possibility it could have forged any elements much heavier than helium.

Gamow had looked to the big bang because the received wisdom in the 1930s was that stars were neither dense enough nor hot enough to support element-building. However, by the 1940s, Fred Hoyle and Ray Lyttleton had recognised that, as stars devoured the nuclear fuel in their cores, they would grow progressively more non-uniform in composition, and that this in turn would make them denser and hotter than Arthur Eddington had ever believed possible.

The pendulum, having swung from the stars to the big bang as the site of the magic furnace, began to swing back to the stars again.

The evidence that atoms were made in stars was more than merely theoretical. By the early 1950s, the idea was receiving support from observations of real stars. In 1951, two American astronomers, Lawrence Aller and Joseph Chamberlain, had found that heavy elements were considerably more common in some stars than in others. On the face of it, this seemed an unremarkable finding. However, the discovery took on a whole new significance when interpreted in the light of an earlier one made by Walter Baade.

During the war, Baade had taken full advantage of the blacked-out skies over Los Angeles to probe the crowded starfields of the Milky Way with the giant 100-inch Hooker telescope. By 1944, the year of his meeting with Fred Hoyle at the foot of Mount Wilson, he had concluded that the galaxy contains two fundamentally different classes of stars, which he christened Population I and Population II.

Stars of Population I are dominated by blue and hot stars. These are found in the Milky Way's 'spiral arms', the filamentary star-lanes which coil themselves around the centre of the galaxy, giving it its characteristic whirlpool shape. Stars of Population II, on the other hand, are red and cool, and are found in the central region of the Milky Way. Aller and Chamberlain for their part had discovered that stars of Population I were richer in heavy elements than stars of Population II.

At first, Baade had no idea what was causing the difference between the two classes of star. However, by 1953, he had become convinced it was age. Population I stars are young. They were born relatively recently in the 'stellar nurseries' of the galaxy's spiral arms. Consequently, the most massive stars among them – those that are hot and blue and race through their lives at breakneck speed – have not had time to burn out and die.* Population II stars, on the other hand, are the galaxy's oldest stars. They were born many billions of years ago, when the Milky Way was young. Consequently, the most massive, blue stars among them long ago burnt themselves out, leaving behind only cool, predominantly red stars.

Now Aller and Chamberlain's discovery could be restated. The observation that stars of Population I were richer in heavy elements than stars of Population II was equivalent to the observation that young stars were richer in heavy elements than old stars. This was exactly what would be expected if elements were made in stars. After all, when a star came to the end of its life and either exploded violently or gently puffed off its outermost layers, some of the heavy elements that it had built up during its lifetime would be scattered to the winds of space. There they would mingle with the gas floating between the stars and enrich it with heavy elements. Since such interstellar gas was the raw material out of which new stars were born, each successive generation of stars would start out its life richer in heavy elements than its predecessors.

But although Aller and Chamberlain's discovery bolstered the case for

---

* Our sun, for example, is a young Population I star.

atoms being made in stars, the case was still far from being proven. The problem was that element-building in stars ran into the same obstacle as element-building in the big bang. Nature had chosen not to provide any stable atomic nucleus of mass 5 or mass 8. This blocked the route to making heavier elements, since the two obvious ways of going beyond helium involved adding an additional nuclear building block to make either helium-5 or lithium-5, or else adding another helium nucleus to make beryllium-8. None of these nuclei could survive for more than an instant before falling apart.

All was not quite lost, however. Stars had several crucial advantages over the big bang. First, in the cores of stars material is squeezed together far more tightly than it was when element-building was possible in the big bang. Nuclear reactions happen more quickly at high density because nuclei collide more often when they are close together. A second advantage stars have over the big bang is time. The fireball of the big bang expanded and cooled at such colossal speed that the conditions for element-building existed for only the first 10 minutes or so of the universe's life. Stars, on the other hand, stay dense and hot for millions or even billions of years. There is plenty of time for even quite rare nuclear processes to occur. One such process is the coming together of three helium nuclei simultaneously.

## THE TRIPLE-ALPHA PROCESS

The most likely outcome of such an encounter is that the nuclei simply cannon off each other. However, there is a small chance that at the instant of closest approach the nuclear force might snap together the three nuclei to make a nucleus of carbon-12. This 'triple-alpha' process, as it was christened, is very rare. Nevertheless, it held out the hope of leapfrogging the troublesome beryllium barrier.

The hope, unfortunately, was a rather slim one. Helium nuclei in the heart of a red giant would be flying about at tremendous speed. Encounters between them would be extraordinarily brief. In fact, three nuclei passing in the night would have barely a thousand million million millionth of a second to react together to make carbon. The chance of a nuclear reaction taking place in such a tiny interval of time was minuscule. The triple-alpha process — the great white hope of stellar nucleosynthesis — simply could not work.[*]

---

[*] Nucleosynthesis is the technical term astronomers and nuclear physicists use for the building of atoms.

However, in 1952, a young researcher at New York's Cornell University took a fresh look at the process. Ed Salpeter was a nuclear physicist interested in nuclear reactions that might be important in astronomy. He began investigating the triple-alpha process by looking at collisions not between three helium nuclei, but between two. Such collisions would appear to lead nowhere. After all, the only conceivable product of two helium nuclei was beryllium-8 and that was unstable. However, beryllium-8 was not so unstable that it could not exist at all. Its life expectancy was merely extremely short. Once formed, it survived for a fraction of a second before breaking apart once more into two helium nuclei.

The implication of all this for a red giant was that there were always some nuclei of beryllium-8 in its helium core – nuclei which had been freshly minted but which hadn't yet self-destructed. It was rather like a firework display in which each individual firework has only a fleeting instant of glory, but in which the overwhelming number of fireworks ensures that at any instant there are always some exploding in the sky.

The average lifespan of beryllium-8 is a hundred thousand million millionth of a second. However, Salpeter realised that even this short interval was ten thousand times longer than the time that two helium nuclei spent together as they rocketed past each other deep in the core of a red giant. So although, by human standards, beryllium-8 survives for the merest blink of an eye, by the standards of the submicroscopic world it loiters around for an eternity – a sitting duck for any helium nucleus that happens to fly by.

In fact, a beryllium-8 nucleus remains in the firing line long enough for ten thousand potential encounters with a helium nucleus. Consequently, a two-step process in which a pair of helium nuclei collide and stick to form beryllium-8, and the beryllium-8 is only later struck by a third helium nucleus to make a nucleus of carbon-12, is ten thousand times more likely to occur than the single-step process in which three helium nuclei come together simultaneously.

Of course, how quickly the two-step triple-alpha process can manufacture carbon depends on how much beryllium-8 is hanging around for target practice. Calculating this is not easy: after all, beryllium-8 is constantly being created and destroyed in the helium core of a red giant. Nevertheless, if the temperature and density are high enough, a situation will be established in which the forces of creation and destruction are exquisitely balanced. In such a 'steady state', the concentration of beryllium-8 will remain constant, just as the level of a swimming pool stays constant if water is pumped in as fast as it is pumped out.

The beryllium-8 concentration can be calculated from a knowledge of the temperature and density in the core of a red giant. Salpeter assumed that the temperature was about 100 million degrees and that the density was about 100,000 times that of water. From this, he deduced that there should on average be only one nucleus of beryllium-8 for every billion nuclei of helium.

It was an incredibly tiny concentration. However, it might just be large enough to enable the triple-alpha process to make carbon at a reasonable rate in the heart of a red giant star. At last, there appeared to be a way past the beryllium barrier. The door to building all the elements in nature in stars seemed open.

## ACADEMIC FRUSTRATIONS

On the other side of the Atlantic, the possibility of making carbon directly from helium by a two-step process had also occurred to Fred Hoyle. In fact, it had occurred to him several years earlier in 1948. However, he had been prevented from investigating the idea by a delicate matter of academic etiquette.

Hoyle had given the problem to a PhD student he was supervising. Unfortunately, two-thirds of the way through the calculations, the student had got fed up and abandoned the whole thing. This should have been no more than a minor irritation to Hoyle. After all, he only needed to pick up the calculations and finish them himself. Unfortunately, it was not quite as simple as that. The disaffected student had failed to cancel his PhD registration at Cambridge. In such circumstances, university etiquette required that Hoyle not touch the problem until either the student's three-year PhD registration had lapsed or the student had completed the university's PhD requirements in some other topic.

Prevented from pursuing his idea, Hoyle became thoroughly frustrated with the Cambridge system. That frustration reached a pitch in 1952, when he read a paper by Salpeter on the synthesis of carbon from helium via the triple-alpha process.

Hoyle's frustration with Cambridge guaranteed that, when an invitation to work in the United States came his way in late 1952, he did not hesitate to accept it. There he would take a crucial step beyond Salpeter and, in the process, make one of the most outrageous predictions in the history of science.

# BREAKFAST IN AMERICA

The American invitation came about by accident. In 1952, Hoyle had gone to Rome for a meeting of the International Astronomical Union, the governing body of world astronomy. Walter Baade was chairing the IAU's session on galaxies. However, he had overlooked the need for a secretary to take down the commission's minutes. Since Hoyle was sitting in the audience, Baade asked him to help. As Hoyle scribbled down the minutes that day, he had not the slightest inkling that they would turn out to be of crucial importance in ensuring that his friend received the credit he deserved for a major astronomical discovery.

This discovery was that the universe was almost twice as old as everyone had previously thought. Instead of being in existence for 2 billion years, it had in fact been around for 3.6 billion years. The discovery was indisputably Baade's. Nevertheless, an outrageous attempt was made to rob him of the glory when the Rome meeting was over.

Baade's own naïveté was undoubtedly a contributory factor. He had announced his new age for the universe in front of an audience rather than in the pages of an astronomical journal. A group of unscrupulous astronomers quickly published the figure of 3.6 billion years themselves, claiming that they had arrived at it before Baade. Hoyle's minutes, were used as evidence against their claim, and ensured that the German astronomer received the glory he deserved.

Baade was clearly grateful because Hoyle received an invitation to spend the first three months of 1953 at the California Institute of Technology in Pasadena. Baade was on the steering committee of both Caltech and Mount Wilson Observatory.

The United States boasted a standard of living far higher than that of post-war Europe, so the idea of working there was extremely attractive. Hoyle accepted the invitation and, at the end of the autumn term of 1952, left Cambridge for Caltech.

The sweet scent of oranges was in the air as he strolled about the pretty campus with its Spanish-style buildings, olive trees and neat, sprinkled lawns. To the north, towering above the campus, was the lofty summit of Mount Wilson, where eight years before while the war still rumbled on he had spent such a memorable weekend. The dome of the 100-inch Hooker telescope flashed silver in the Californian sunshine.

As he strolled along the Olive Walk, he had high hopes for his short stay in California. They were hopes which would be more than fulfilled.

# TROUBLE FOR THE TWO-STEP TRIPLE-ALPHA

Hoyle's first task was to give a series of lectures for the combined astronomers of Caltech and nearby Mount Wilson. As part of his preparation for the lectures, he took a close look at the triple-alpha process. It wasn't long before he began to wonder whether Salpeter's claims for the process hadn't been just a little over-optimistic.

According to Hoyle's calculations, there simply had not been enough time since the big bang for the triple-alpha process to have made all the carbon in the universe. The problem lay with the second step of the process – the one in which the third helium nucleus collided and stuck with beryllium-8 to make carbon-12. The nuclear reaction which snapped together beryllium-8 and helium-4 was simply too slow.

As Salpeter had calculated, the core of a red giant contained only about one beryllium-8 nucleus for every billion nuclei of helium. To compensate for this imbalance, and for carbon to be made at an adequate rate, the nuclear reaction which glued together beryllium-8 and helium-4 had to occur in a relatively large percentage of encounters between the two nuclei. However, when Hoyle examined the process in detail, he discovered that the nuclear reaction which snapped together beryllium-8 and helium-4 was far too rare. The triple-alpha process, as envisioned by Salpeter, was quite incapable of making the large quantities of carbon that evidently exist in nature.

Unfortunately, nobody had come up with any other plausible way to bypass the troublesome beryllium barrier. If the triple-alpha process was not the solution, astronomers would have no clue how nature managed to assemble the huge majority of the elements in the universe. With this in mind, Hoyle began to look for a way for it to work.

# A VERY SPECIAL STATE OF CARBON

The one-in-a-billion concentration of beryllium-8 could not be varied. It was fixed by the lifetime of beryllium-8, and by the density and temperature in the core of a red giant. Hoyle therefore focused his attention on the nuclear reaction between beryllium-8 and helium-4. Was there any chance that it could be speeded up?

There was. However, it required the nucleus of carbon-12 to have a very special property. It needed to have an energy almost exactly equal to the combined energy of beryllium-8 and helium-4 at the typical temperature deep inside a red giant. If, by some fluke, there was such a match, then the beryllium-8 and helium-4 would be processed very readily into carbon-12. The nuclear reaction, in the technical jargon, would be 'resonant'.

The energy of a carbon-12 nucleus, like the energy of any atomic nucleus, is determined by the behaviour of its constituents – the proton and neutron building blocks. If they jostle about frantically inside, then the nucleus possesses more energy than if they move about sluggishly. The only restriction is imposed by the laws of quantum theory. They permit only a limited number of 'energy states', much the way they do with an electron in an atom. In common with an atom, the lowest energy state is called the 'ground state' and all the others 'excited states'. In view of all this, Hoyle's condition for speeding up the nuclear reaction between beryllium-8 and helium-4 could be expressed more precisely: carbon-12 had to possess an excited state with an energy very close to the combined energy of the beryllium-8 and helium-4 inside a red giant.

Hoyle calculated the sum and found the energy of the desired state. It was precisely 7.65 megaelectronvolts (MeV) above the ground state. If carbon had such an energy level, the beryllium barrier would be shattered, and the way to the building of all the heavy elements in nature would be open. It seemed a perfect solution. However, there was one slight snag. According to all the experiments, there was no excited energy state of carbon at 7.65 MeV.

Hoyle, however, had complete faith in his logic. He refused to be deterred by the small matter of experimental results: the 7.65 MeV state simply had to exist. If it did not, Hoyle reasoned, the universe would contain no carbon. And if there was no carbon, there would be no human beings, since human beings, in common with all other living things, are carbon-based organisms. What Hoyle was saying – and nobody had ever used logic as outrageous as this before – was that the mere fact that he was alive and pondering the question of carbon production in stars was proof that carbon-12 had to have an excited state at 7.65 MeV.

Eventually, Hoyle decided to go and talk to one of Caltech's experimental nuclear physicists. The man he called on was Willy Fowler.

## THE MOST OUTRAGEOUS PREDICTION IN SCIENCE

Fowler was the man who had measured the speed with which the carbon-nitrogen cycle converted hydrogen into helium in stars, thereby proving that, in the sun at least, the conversion had to be achieved by a competing proton-proton chain. For his pioneering work on the carbon-nitrogen cycle, Fowler was considered the founding father of a new science – experimental nuclear astrophysics.

Contact with astronomers at both Caltech and nearby Mount Wilson had made him unusually receptive to their ideas. However, nothing had prepared him for the outrageous claim made by Fred Hoyle when he turned up in the Kellogg Radiation Laboratory in the spring of 1953.

Atomic nuclei are hideously complicated. The prediction of any of their properties requires knowledge of the precise manner in which all of their protons and neutrons behave. However, the mathematical techniques to find out such a thing simply did not exist. The most complicated system whose behaviour physicists could precisely predict was a 'two-body system', such as the moon going round the earth. The 'many-body system' of an atomic nucleus was way beyond their capabilities.

Yet here, sitting in Fowler's office, was an astronomer claiming he could do what no nuclear physicist in the world could do: predict a precise energy state of an atomic nucleus. What made his claim even more outrageous was that the prediction was based not on considerations of nuclear physics, but on the argument that we, as human beings, exist, therefore carbon-12 must possess an unknown energy state at 7.65 MeV.

It was highly probable that Hoyle was wrong. On the other hand, if he was right, the consequences for element-building in stars would be profound. It was this consideration – not to mention a sneaking admiration for a man with the bare-faced nerve to walk into Kellogg with such a preposterous claim – that prevented Fowler from showing Hoyle the door.

This was the smartest move of Fowler's life – one which would ultimately earn him the Nobel prize. He listened to what the English astronomer had to say, then rounded up the members of his small research group. Sitting in Fowler's office, surrounded by down-to-earth experimentalists, Hoyle was acutely aware that his scientific reputation was now on the line. He repeated his extraordinary argument for the existence of the 7.65 MeV excited state of carbon-12, and asked whether there was any possibility that the experiments to date had somehow missed it.

Much of the technical discussion that followed went over Hoyle's head. Eventually, however, a consensus was reached among Fowler's colleagues. Yes, if the excited state of carbon-12 had certain very special properties – termed 'even parity' and 'zero spin' – it was just possible that experiments might have missed it.

Hoyle's idea had passed its first major test, and his reputation was still intact. Most importantly, he had piqued the interest of Kellogg's nuclear physicists. Now they were aware of the possibility that an energy state of carbon-12 had been overlooked, they had no choice but to set up an experiment to look for it.

For 10 days, as the experiment proceeded, Hoyle was on tenterhooks. Each afternoon, he crept down into the bowels of Kellogg, where Fowler's colleague Ward Whaling and his team beavered away amid a jungle of power cables, transformers and diving-bell-like chambers in which atomic nuclei were fired at each other. And each afternoon, he crept back up again into the painfully bright Californian sunshine, relieved that his idea had survived one more day without being disproved. On the tenth day, however, Hoyle was met by Whaling. He pumped Hoyle's hand and gushed his congratulations. The experiment had succeeded. Hoyle's prediction had been borne out. Quite unbelievably, there was an energy state of carbon-12 very close to 7.65 MeV.*

It was the most amazing result that Fowler had ever seen emerge from Kellogg. Never had he believed that Hoyle's outrageous prediction would actually be proved right. But it had been – quite spectacularly. But what compounded Fowler's amazement was the manner of Hoyle's prediction. He had predicted the 7.65 MeV energy state of carbon-12 using a so-called anthropic argument: that the state had to exist because, if it did not, neither could carbon-based creatures such as human beings. To this day, Hoyle is the only person to have made a successful prediction from an anthropic argument in advance of an experiment.

## COSMIC COINCIDENCES

When the euphoria had finally faded, however, Hoyle was left to ponder the bizarre picture that had been suddenly revealed to him. It was now clear that the existence of the 20 or so heavy elements essential for life – elements like carbon and oxygen and iron – was dependent on a series of quite remarkable coincidences. Those coincidences involved the 'fine-tuning' of the properties of at least three separate atomic nuclei.

The first of these involves beryllium-8. For years, the nucleus's instability had appeared to be a major obstacle to the forging of carbon and of all elements heavier than carbon. Now, Hoyle saw that the fleeting existence of the nucleus was actually a blessing in disguise. If nature had chosen to make beryllium-8 stable, no sooner had a star

---

* Today's precise measurements put it at exactly 7.6549 MeV.

become hot enough to glue together helium nuclei than nuclear reactions would instantly transform all its helium into carbon. The star, unable to cope with the sudden and violent release of nuclear energy, would blow itself apart. The destruction of the star would rule out the possibility of making elements that were essential to life such as calcium, magnesium and iron.

The second nucleus whose properties appeared to have been fine-tuned was carbon-12. For the triple-alpha process to work and for human beings to exist, carbon-12 had to possess an excited state with an energy almost exactly equal to the combined energy of a beryllium-8 nucleus and a helium-4 nucleus. The combined energy of beryllium-8 and helium-4 is actually 7.3667 MeV – just below the 7.6549 MeV energy state in carbon-12. However, the temperature in a red giant star is high enough to boost the energy of motion of the two nuclei to just above the critical threshold of 7.6549 MeV. For carbon to be created, the excited energy state of carbon-12 had to be just right and so did the temperature inside a red giant. It was a coincidence to strain the bounds of credibility.

But this was not the end of nature's fine-tuning. Once a carbon-12 nucleus has formed in the core of a red giant, it is bound to be struck by another helium-4 nucleus. If, during such a collision, there is enough time for a nuclear reaction to glue together carbon-12 and helium-4, then as fast as carbon-12 is formed it will be transformed into oxygen-16. All the good done by the triple-alpha process would be undone, and the universe's carbon supply would be seriously depleted. Our very existence as carbon-based creatures, therefore, requires that the nuclear reaction between carbon-12 and helium-4 is not too fast.

On the other hand, if, during a collision between carbon-12 and helium-4, there is insufficient time for a nuclear reaction to glue together the two nuclei, then very little oxygen-16 will be made, and oxygen stocks will be too low to support life.

A delicate balancing trick is required if neither oxygen nor carbon is to be over-abundant in the universe, at the expense of the other. Just as the nuclear reaction between beryllium-8 and helium-4 has to proceed slowly but not too slowly, so too must the reaction between carbon-12 and helium-4.

This is the case, Hoyle was able to show, only if the reaction between carbon-12 and helium-4 is not resonant – that is, if oxygen-16 does not possess an excited state with an energy equal to the combined energy of a carbon-12 nucleus and a helium-4 nucleus, at the temperature inside a red giant.

The combined energy of a nucleus of carbon-12 and of helium-4 is 7.1616 MeV. The energy state of oxygen-16 is extremely close to the danger level, at 7.1187 MeV. However, by an amazing piece of good fortune, this energy state is just below the energy of carbon-12 plus helium-4. Since the high temperature inside a star can only add energy of motion to the reacting nuclei, and not subtract it, there was no possibility that the combined energy of carbon-12 and helium-4 could be made to equal 7.1187 MeV.

Human beings had escaped by the skin of their teeth. Had the energy state of oxygen-16 been a fraction above 7.1616 MeV instead of a fraction below, the nuclear reaction between carbon-12 and helium-4 would be resonant and all the carbon made inside stars would turn instantly into oxygen-16. There would be hardly any carbon in the universe. Life is possible because nature has fine-tuned the properties of three different atomic nuclei. Beryllium-8 is unusually long-lived for an unstable nucleus. Carbon-12 possesses the exact energy state needed to promote its production. And oxygen-16 lacks an energy state that would promote its production at the expense of carbon.

But what do all these coincidences mean? One explanation was that the universe had actually been designed to permit the evolution of life. This was, of course, the religious view. Hoyle, as a scientist, however, preferred to consider more prosaic explanations.

One possibility was that beyond our universe are other universes, each of which had its own, slightly different, laws of physics. In some universes, for instance, the nuclear force might be stronger or the gravitational force weaker than in ours; permutations are endless. If there were indeed a multitude of such universes, then it is clear that life would arise only in those where the nuclear and electrical forces conspire to give beryllium-8, carbon-12 and oxygen-16 the very special properties that they possess in our universe. In the huge majority of universes, these nuclei would have properties other than the desired ones and, consequently, there would be no intelligent life forms to remark on any coincidences. This curiously topsy-turvy logic was known as the 'anthropic principle'. In short, it maintained that things are the way they are because if they were not the way they are we would not be here to notice the fact!

The remarkable series of nuclear coincidences first recognised by Hoyle was to have a profound influence on his life. It convinced him that the universe was geared up for the emergence of living organisms. Life had arisen on earth not because of some zillion-to-one accident but because it was truly a cosmic phenomenon.

# THE ALPHA PROCESS

By showing how helium could be transformed into carbon inside a red giant, Hoyle had removed the last major barrier to understanding how the elements were built up inside stars. It was now possible to see a way of making several of the elements that were found on earth – elements such as oxygen, which could be produced by a helium nucleus slamming violently enough into a carbon nucleus that it stuck fast, and neon, which could be created by a helium nucleus burying itself inside a nucleus of oxygen.

This build-up process in which helium nuclei, or alpha particles, were glued to a nucleus one at a time was dubbed the 'alpha process'. In addition to making oxygen-16 and neon-20, it is capable of making nuclei such as magnesium-24, silicon-28, argon-36 and calcium-40. All of these are noticeably more common than nuclei which are either slightly larger or smaller – evidence which appeared to support Hoyle's contention that they were products of the alpha process.

By the time Hoyle left Caltech for England in March 1953, he had worked out the full details of the first two steps in the alpha process – the conversion of carbon to oxygen, and the conversion of oxygen to neon. He had also drafted an ambitious paper which would appear in the *Astrophysical Journal* the following year under the title 'I: The Synthesis of the Elements from Carbon to Nickel'.

Hoyle had now identified two distinct processes capable of building up heavy elements: the 'equilibrium' process which made the iron-peak elements, probably in the interior of an exploding supernova, and the 'alpha' process, which made elements such as oxygen, sulphur, silicon and calcium, probably in massive, hot stars. Once even a few of these heavy elements had been made inside a star, it opened a whole host of other element-building possibilities because nuclei might react with each other. For instance, in a star which had already transformed much of its core into carbon, two carbon nuclei might collide and produce magnesium-24.

Of course, magnesium-24 can also be made from neon-20 by the alpha process. In fact, this is the more likely route since the electrical repulsion between a helium nucleus and a nucleus of neon is a lot weaker than that between two nuclei of carbon. Nevertheless, the mere fact that magnesium-24 can be made by more than one route demonstrates the complex range of reactions which can take place inside a star.

Despite all the possible permutations of nuclear reactions, however, there was still no conceivable way of synthesising many of the elements in nature. For instance, it was impossible to see how to make very heavy elements such as gold, tin and uranium. All of these are heavier than iron and would have to be made by adding a succession of nuclear building blocks to iron or other iron-peak elements.

This posed formidable problems. The nuclei of the iron-peak elements are at the very bottom of the curve of binding energy, and so are the most strongly bound of all nuclei. Turning them into heavier nuclei is equivalent to driving them uphill, requiring an input of energy. The trouble is that the very stability of stars depends on nuclear reactions that give out energy, thus generating the heat to defy gravity. It was hard to see how stars could tolerate vampire-like nuclear reactions in which their heat-energy was actually sucked out.

But there was another, more serious difficulty in forging very heavy elements. It requires the gluing together of large nuclei. And large nuclei, on account of their great electrical charges, repel each other with a vengeance. The only way to bring them close enough to fuse is to slam them together at speeds approaching that of light itself. However, such speeds are never reached by large nuclei, even in the hottest stellar interiors imaginable.

Electrical repulsion placed a fundamental limit on element-building. It ruled out the possibility of making the heaviest atoms in nature by banging together charged nuclei. There simply had to be a third element-building process in addition to the equilibrium and the alpha processes.

# 14

# The Crucibles of Stars

## HOW WE FINALLY DISCOVERED THE MAIN PROCESSES THAT MAKE THE ATOMS.

It is the stars,
The stars above us, govern our conditions.

William Shakespeare

Fred Hoyle was well aware of the impossibility of making the very heaviest elements by slamming together pairs of charged nuclei inside stars. At some future date he intended to write a follow-up paper to his paper on forging the medium-weight elements, in which he would investigate an entirely different class of element-building reactions: those that involve neutrons.

A neutron, on account of its lack of electrical charge, is not repelled by an atomic nucleus. It can slip right through the walls of the nuclear fortress and become ensnared by the powerful nuclear force. It was by a series of such 'neutron captures' that Hoyle envisaged a 'seed' nucleus – for instance, iron – being built up into heavier and heavier nuclei.

In fact, this neutron build-up scheme was very similar to the one considered by Carl-Friedrich von Weizsäcker in the late 1930s. The German nuclear physicist had abandoned the idea, however, when told by astronomers that the extraordinary range of temperatures and densities required for the scheme simply did not exist inside stars. (This had of course been George Gamow's cue to look for an alternative furnace for atom-building and invent the hot big bang.) Between the 1930s and 1950s, however, there had been a revolution in our understanding of stars. Arthur Eddington's view of stellar interiors as being essentially all alike, with central temperatures never much higher than the 10 million degrees or so at the heart of the sun, was abandoned. In its place was substituted a much more diverse picture, with the most massive ones becoming ever more dense and hot as they burnt up first one nuclear fuel, then another. In such extreme and varied conditions all manner of element-building processes might go on, including ones that involved the capture of neutrons.

Hoyle's hopes for the neutron build-up process were much more modest than von Weizsäcker's. Unlike his German predecessor, he did not expect the process to be responsible for all the atoms in the universe, only those that were passed over by the alpha and equilibrium processes – essentially, the very heaviest elements. For any neutron build-up scheme the first requirement was obviously neutrons. Hoyle's most pressing problem was therefore to identify a plausible source inside stars.

## A SOURCE OF FREE NEUTRONS

Maintaining a constant supply of free neutrons inside a star is like maintaining a constant supply of lemmings on a clifftop. Neutrons are born with an irresistible death wish and, after roughly 10 minutes outside an atomic nucleus, abruptly disintegrate. Furthermore, the interiors of stars contain a large number of neutron-hungry nuclei, any one of which might gobble up a free neutron long before its allotted 10 minutes.

The nasty, brutish and short life of the average neutron necessitated finding a nuclear process that replenished neutrons as fast as they either were gobbled up or committed suicide. Off the top of his head, Hoyle could think of two obvious possibilities: 'carbon burning', which converts carbon into oxygen, and 'oxygen burning', which turns oxygen into neon. Each process, after running its course, leaves behind a single surplus-to-requirements neutron. Such leftover neutrons are ideal for driving the neutron build-up process; there seemed no need for Hoyle to keep looking. However, it turned out that there was a much better source of neutrons inside stars. It was discovered not by Hoyle, but by a young Canadian called Alastair Cameron.

Cameron was a nuclear physicist at Chalk River near Montreal. He had become interested in the problem of building up atoms in stars after reading a short article in an American popular science magazine in 1952. It described the work of Paul Merrill, an astronomer at the Mount Wilson and Palomar Observatories in California. Merrill had discovered an obscure element called technetium in the light of a dozen red giant stars. This was a great surprise because technetium is highly unstable;

even its longest-lived isotope self-destructs with a half-life of a mere 4.2 million years. In other words, any celestial body much older than 4.2 million years should long ago have lost any technetium it was born with. This is certainly true for the earth, which has not the slightest trace of naturally occurring technetium. And it should also be true for red giants, since such stars are typically even more ancient than the earth.

The surprise was that Merrill's red giants very definitely did contain technetium and in amounts large enough to leave the element's spectral fingerprint in the light of each star. There could be only one explanation. The stars must have made the technetium recently. And, since it was highly unlikely that Merrill had happened on them in the immediate aftermath of a technetium-forging episode, the stars must be making the element continuously.

It was a discovery to top even Aller and Chamberlain's. Their finding – that young stars were richer in heavy elements than old ones – merely indicated that at some time in the distant past atom-building had taken place inside stars. Merrill's discovery proved that stars were making atoms now. Cameron knew that the creation of technetium must require a source of neutrons. The magazine article also stated that Merrill had found the elements barium and zirconium in his stars; their manufacture also needed neutrons. But where were the neutrons coming from?

Like Hoyle, Cameron began to look for a nuclear process that could make neutrons faster than they could disintegrate or be gobbled up by predatory nuclei. Before long he came to a process that seemed to fit the bill. It was the fusing of a nucleus of helium-4 and a nucleus of carbon-13, a heavy isotope of carbon, to make a nucleus of oxygen-16. After the nuclear dust had settled, a single neutron would be left over.

Since the reaction requires helium, it would obviously have to go on in a part of a red giant where there was plenty of helium. This turned out to be fortunate, since any region rich in helium would be correspondingly depleted in hydrogen, the raw material from which helium was made. Hydrogen is one of the principal neutron gobblers in stars.

Everything fitted together. The build-up of heavy elements – including the technetium, barium and zirconium seen by Merrill in his red giants – required only that the neutrons unleashed in Cameron's process collide and stick to seed nuclei like iron. However, there was one problem. Whether or not Cameron's process could generate enough neutrons to drive the neutron build-up process depended on how much carbon-13 there was in a star. After all, it was the reaction between carbon-13 and helium-4 that produced the all-important leftover neutron. The trouble was that carbon-13 was an extremely rare form of carbon.

What Cameron needed to find was a process that could boost the abundance of the isotope inside a star. Fortunately, such a process existed: the carbon–nitrogen cycle. It was responsible for turning hydrogen into helium in the most massive stars. And, as a by-product, it created small amounts of carbon-13. Carbon-13 would be unable to collide and stick with helium-4 right away. The temperature was simply too low inside a star undergoing the carbon–nitrogen cycle. Somehow, the carbon-13 had to survive until a later, hotter stage in the star's life. Granted that this was possible, however, the critical neutron-releasing reaction would go ahead.

It seemed that Cameron had constructed a house of cards. This may well have been the opinion of the two referees who rejected the idea when Cameron submitted a paper to the *Astrophysical Journal*. In normal circumstances, this would have killed the paper stone dead. However, the editor of the journal had a sneaking suspicion that the paper had some merit, and decided to seek a third opinion. The person he turned to for that opinion was Fred Hoyle.

Choosing Hoyle was no accident. The editor of the *Astrophysical Journal* was the Indian astrophysicist Subrahmanyan Chandrasekhar, who had worked at Cambridge in England before emigrating to the United States. He knew Hoyle personally, and he also knew of Hoyle's pioneering work on element-building in stars. Cameron's paper reached Hoyle shortly after he had returned from America. As soon as he began reading it, he realised that it contained a brilliant idea. Here was a far more plausible source of neutrons than the carbon-burning and oxygen-burning sources he had chosen.

Hoyle's only concern was whether the carbon–nitrogen cycle could really generate sufficient carbon-13 for everything to work. Calculations he had carried out with the Princeton theorist Martin Schwarzschild before returning to England had a bearing on this. They had revealed the complex situation that can develop towards the end of a star's life, when several different nuclear fuels might be burning, each in a different part of a star. It can be so complicated, in fact, that it was impossible to rule out Cameron's neutron-generating scheme. Hoyle's advice to Chandrasekhar was therefore to publish the Canadian physicist's paper. Today, Cameron's carbon-13 source is widely believed to be the neutron source for element-building in all red giants with masses less than eight times that of the sun.[*]

---

[*] In red 'supergiant' stars with a mass greater than eight times that of the sun, the source

With the identification of a plausible source of neutrons inside stars, the next problem was to determine whether the neutron build-up scheme was capable of making the heaviest elements in the same relative proportions seen in nature. Cameron set about tackling the problem alone, Hoyle as part of a formidable 'gang of four'. The other members were Willy Fowler and a husband–and–wife team, Geoffrey and Margaret Burbidge.

## THE S-PROCESS

Like Hoyle, Geoffrey Burbidge had started out as a theoretical physicist. He had got into astronomy, however, when he had married an astronomer: Margaret Burbidge, who became the first woman Astronomer Royal in the post's 300-year history. The Burbidges arrived in Cambridge in 1954 and in the autumn of that year Geoffrey gave a talk on their joint research. In the audience was Willy Fowler. Fowler was in Cambridge because he was owed a sabbatical year by Caltech and had decided to spend it learning as much astronomy as possible, at the university that employed Hoyle.

The Burbidges had been measuring the abundances of different elements in stars. In the course of this work, they had stumbled on some extremely peculiar stars in which elements heavier than iron were far more common than was normal. On hearing this news, Fowler pricked up his ears. When the talk was over, Fowler caught up with Geoffrey Burbidge. He introduced himself as an experimental nuclear physicist, and pointed out that the unusual element abundances might be explained if nuclear processes inside the stars were actually manufacturing the heavy elements. It was the start of an extremely fruitful collaboration.

Soon Fowler and the Burbidges were going over the same mental territory as Hoyle and Cameron. It was obvious that the very heavy elements the Burbidges were seeing in their stars could not have been made by slamming together charged nuclei. They would repel each other too strongly. The only sensible alternative was that the heavy elements had been built up from a seed nucleus like iron by adding neutrons one at a time.

It was an extremely slow business. Even with a scheme like

---

of free neutrons is thought to be the reaction between neon-22, a rare isotope of neon, and helium. When the two nuclei meet and stick, the result is magnesium-25 plus a left-over neutron. The neon-22 neutron source was identified in 1960 by Alastair Cameron.

Cameron's generating the neutrons, the particles would be enormously rare. After being struck by one neutron, a nucleus would typically have to wait hundreds of thousands of years before it was struck by another. In recognition of this fact, Fowler and the Burbidges christened the neutron build-up scheme the slow or s-process.

The s-process could make elements heavier than iron by a sequence of neutron captures, punctuated by beta decays when any unstable nucleus which was formed turned one of its neutrons into a proton. Precisely which elements and in which proportions depended on how long was available. If there was plenty of time, then every nucleus would gobble up neutrons until it reached the point where it could gobble up no more. Everything would therefore be transformed into the heaviest nucleus that the s-process was capable of making, which turned out to be the metal bismuth, with 83 protons jammed into its nucleus.

However, the evidence from the real world was that there was a whole range of s-process elements – not just the heaviest ones but light and medium-weight ones as well. This could only mean that the s-process was often choked off long before it could run its course. The most likely possibility was that sooner or later a red giant became unstable and puffed off its outer layers. All the s-process elements the star had made up to that point in its life would then be expelled into space.

The s-process could make a whole range of heavy elements inside stars, including the technetium, barium and zirconium which Merrill had spotted in his red giants. However, the mere fact that it could make such elements was no proof that it really was making them. Proof was needed that the s-process was capable of making heavy elements in the same relative proportions as seen in nature.

The early calculations of Fowler and the Burbidges were encouraging. They selected a relatively light seed nucleus – neon-20 – and worked out what would happen if neutrons were slowly added, one at a time. One factor which would influence how much of each nucleus is made is how eagerly it snaps up any neutrons which happen to come its way – its so-called neutron-capture 'cross-section'. Using cross-sections measured in the laboratory by other researchers, Fowler and the Burbidges were able to predict which nuclei the s-process should make in large amounts and which in small amounts. When they compared the relative abundances of elements with an odd number of nucleons with the predictions of the s-process, they found a near-perfect match. It was the first solid evidence that the process had played a role in forging the elements.

There were no reliable neutron capture cross-sections for heavy nuclei, so the team had to stop their calculations at scandium-46. The necessary information was kept under lock and key at the US Atomic Energy Commission and was not declassified until 1956. By that time, however, Fowler and the Burbidges were at Caltech in California, where they were finally joined by Hoyle.

Hoyle had not teamed up with Fowler and the Burbidges immediately because he was burdened with a full teaching load at Cambridge. In 1956, however, he received an invitation to spend more time at Caltech. On his arrival in the spring, he sought out Fowler and the Burbidges. He found them working together in a windowless office deep in the basement of the Kellogg Laboratory. The time was now ripe to nail the problem of the origin of the elements. Not only were neutron–capture cross-sections available, and recently declassified, but accurate estimates of the cosmic abundances of the elements had just been published by American scientists Hans Suess and Harold Urey.

One day Hoyle noticed that there was a definite connection between the neutron–capture cross-section of every heavy nucleus and its abundance in the universe. Heavy elements with small cross-sections were relatively common whereas elements with large cross-sections were relatively rare. It was exactly what would be expected if the elements had been manufactured by the s-process. If a nucleus's neutron–capture cross-section was small, it should wait for a long time before swallowing a neutron and transforming itself into a different nucleus. Consequently, it ought to be common. On the other hand, if a nucleus's neutron–capture cross-section was large, it should snap up a neutron at the slightest opportunity and turn into something else. Consequently, it ought to be rare. Exactly the same logic applies to human populations: there are more adults than there are babies because people spend far more time as adults than they do as babies.

Further evidence came from observations made by Margaret and Geoffrey Burbidge with a telescope on Mount Wilson. One of the stars they examined was a yellow giant called HD 46407 in the constellation of Canis Major, which turned out to be rich in barium, strontium, yttrium and zirconium. Not only were these all elements which the s-process could make, but their relative abundances in the star perfectly matched the theoretical predictions of the s-process.

Despite its success in explaining the origin of many elements, however, the s-process had a rather serious shortcoming. It was not capable of making *all* the elements heavier than iron.

# THE R-PROCESS

At the root of the problem was the slowness of the process. The long gap of 100,000 years or so between neutron captures meant that, if an unstable nucleus was formed, there was normally time for it to beta-decay before the next neutron came along. Since a beta decay always transformed a neutron into a proton, the s-process was quite hopeless at making nuclei which were rich in neutrons.

This shortcoming of neutron build-up schemes had in fact been recognised by von Weizsäcker back in 1938. He had discovered that, although the slow addition of neutrons to a 'seed' nucleus like iron could make elements such as barium and zirconium, elements like thorium and uranium could not be made.

Thorium and uranium are neutron-rich. Making them is impossible while beta decay keeps undoing the good work by turning neutrons into protons. The only way to get around the problem is to make sure that any unstable nuclei formed have no opportunity to beta-decay. In practice, this meant subjecting nuclei to such a ferocious neutron bombardment that any unstable nucleus formed is always hit by another neutron before it has time to decay. Instead of one neutron every 100,000 years or so, this requires a rate of at least one neutron every second.

Von Weizsäcker had concluded that the only way a neutron build-up scheme might have a chance of making all the elements was if neutrons were supplied at two entirely different rates: extremely slowly, to make neutron-poor elements like barium and zirconium, and extremely rapidly, to make neutron-rich elements like thorium and uranium. But how could a single source of neutrons satisfy two such contradictory requirements?

The answer of course was that it couldn't. This was clear to Hoyle from the moment he had started considering the possibility of building heavy elements by adding neutrons. Inside stars, there had to be two distinct processes: a slow addition of neutrons to an initial seed nucleus like iron – the s-process – and a rapid addition to a similar seed nucleus, which Hoyle's team soon dubbed the rapid or r-process.

Two neutron-capture processes were far easier for Hoyle to contemplate than von Weizsäcker. The German physicist, after all, had been looking for a single process to make all the elements in the universe; Hoyle, on the other hand, had early on realised that there was a whole range of nuclear processes.

Hoyle decided to calculate the effect of adding neutrons rapidly to iron. The results were encouraging. Estimates for the abundance of r-process elements agreed with observations. In fact, the r-process would turn out to have produced most of the elements heavier than iron – elements such as gold, silver and platinum. But, just as with the s-process, the big question was: where did the neutrons come from? The source of r-process neutrons would clearly have to be vast in order to pepper every nucleus with a neutron roughly every second.

On earth, the only conceivable source of such a hailstorm of neutrons was a nuclear explosion. In fact, data from an American H-bomb test on Bikini Atoll declassified in the 1950s had revealed the presence of super-heavy nuclei which could only have been created by the intense flux of neutrons – in other words, the r-process.

Hoyle had already made a connection between the explosion of a nuclear weapon and the explosion of a star as a supernova. It had provided him in 1945 with the first solid evidence that element-building went on in stars. It was therefore natural for Hoyle to speculate that supernovae were the places where the r-process went on.

The guess appeared to receive dramatic confirmation from a supernova in a galaxy called IC4812. The supernova had gone off in 1937, and been observed for the two years which followed by Walter Baade. On examining Baade's records, Geoffrey Burbidge noticed that the supernova had faded in a rather intriguing manner. After about 55 days, its brightness had fallen to half its initial level, after another 55 days to a quarter, after a further 55 days to an eighth, and so on. This kind of behaviour was strikingly similar to the decay in activity of a radioactive element with a half-life of 55 days.

As it happened, the bomb test on Bikini Atoll had revealed the presence of an isotope of an artificial element with a half-life of around 60 days. It was known as californium-254. If the r-process could make it in the explosion of a nuclear bomb, why not in the explosion of a supernova? Burbidge therefore made the bold suggestion that it was the decay of californium which was behind the fading of the light in Baade's supernova.

The general idea turned out to be right. Unfortunately, the details were wrong. Astronomers now know that the decaying light of a supernova actually involves two other radioactive isotopes: nickel-56, which has a half-life of 6 days, and cobalt-56, which has a half-life of 77 days. Together, the decay of these two nuclei mimics that of a single radioactive isotope with a half-life of 60 days.

For 18 months, the team worked on the puzzle. By September 1956 they had decided to publish their material in an encyclopedic article.

## 'B-SQUARED-FH'

It was now possible to show how most of the elements in the universe had been made by a series of element-building processes. In their article the team set out and summarised all the processes for the first time.

The first process was of course 'hydrogen-burning', the conversion of the lightest element, hydrogen, into the second lightest element, helium. Occurring at the relatively modest temperature of 15 million degrees at the heart of the sun, it is the power source of the huge majority of stars. The details of hydrogen-burning – either via the proton-proton chain or the carbon-nitrogen cycle – had of course been worked out by Hans Bethe, Charles Critchfield and Carl-Friedrich von Weizsäcker in 1939. Willy Fowler later showed that in the sun the proton-proton chain is the more important. It is responsible for sunlight and all life on earth.

The second element-building process is the triple-alpha process, the sticking together of three helium nuclei to make carbon-12. It occurs at the much higher temperature – 100 million degrees or so – found deep inside red giants, the stars at a later stage of life than the sun which have used up all the hydrogen fuel in their cores. Here the details had been worked out by Ed Salpeter and Fred Hoyle. Once again, Fowler, or at least Fowler's colleagues at Caltech's Kellogg Laboratory, had played a key role in demonstrating the workability of the scheme.

By bypassing the obstacle of beryllium, the triple-alpha process opened the way to forging all the heavy elements in the universe. The most straightforward way is by taking carbon and adding a succession of helium nuclei, or alpha particles. By means of this third element-building scheme, dubbed the alpha process, it is possible to make oxygen-16, neon-20, magnesium-24, silicon-28, argon-36 and calcium-40. It requires even more extreme conditions than the triple-alpha process, with stellar temperatures ranging from 100 million degrees up to a billion degrees.

But the alpha process is very selective, skipping over many nuclei. More seriously, it is constrained by the ever-growing electrical charge on the nuclei it is making. When that charge becomes large, alpha particles are repelled so violently that even in the hottest possible stars, where alpha particles are travelling at tremendously high speeds, they cannot get close enough to a nucleus to stick.

Almost the only way to create really heavy nuclei, not to mention some of the nuclei missed out by the alpha process, is to add particles which are unaffected by the electrical charge of a nucleus. In practice, this means neutrons. The fourth and fifth element-building processes are therefore neutron build-up schemes. Both start with an appropriate 'seed' – either an 'iron-group' element or a light element – and proceed to pelt it with neutrons. In the case of the s-process, this takes place so mind-bogglingly slowly – only about one neutron every 100,000 years or so – that there is usually plenty of time for the decay of any unstable nucleus which forms before the next neutron comes along. In the case of the r-process, on the other hand, the rate is so fast – one neutron approximately every second – that there is no time for an unstable nucleus to decay before the arrival of the next neutron.

The s-process, operating inside red giant stars, produces a whole range of heavy nuclei such as technetium, barium and zirconium. However, it is useless at building nuclei which are rich in neutrons, and it cannot make nuclei heavier than bismuth-209, with 83 protons in its nucleus. This situation is remedied by the r-process, which most probably goes on in the fury of supernova explosions. It creates a different range of nuclei and they include some of the heaviest elements known, such as thorium, uranium and californium.

But the s- and r-processes, although they can make most heavy nuclei, cannot make them all. Some heavy elements possess rare isotopes with relatively few neutrons: for example, tin-112, platinum-190 and mercury-196. The only way to explain these proton-rich elements was to postulate a sixth element-building process, capable of adding protons, rather than neutrons, to a seed nucleus. Hoyle and his colleagues called it the proton-capture process, or p-process. It requires a hydrogen-rich medium and a temperature of over a billion degrees, which means it probably occurs in supernova explosions.

Each of the first six element-building processes involves assembling elements in a series of steps, taken one at a time. But there is another element-building process which works in quite a different way. It is the equilibrium process, in which the abundances of various elements magically 'freeze out' in the midst of a frenzy of nuclear reactions.

The equilibrium process is the seventh element-building process. Most probably operating in the nuclear melt-down of a supernova explosion, it requires a temperature of more than a billion degrees to work. Its fingerprint, which was stamped indelibly on the abundances of the iron-group elements, had first been spotted by Hoyle in 1945, and was the single piece of evidence that convinced him that stars were the furnaces where atoms were forged, setting him on the road that culminated in his ground-breaking work with Margaret and Geoffrey Burbidge and Willy Fowler.

The paper in which the team laid out the seven nuclear processes operating inside stars and supernovae ran to 104 pages. It was published in 1957 in *Reviews of Modern Physics* – the only journal thick enough to take it – and has ever since been known as 'B-squared-FH', after the initials of its four authors.

'B-squared-FH' was an intellectual tour de force. It showed, conclusively and clearly how, for instance, most of the world's gold, uranium and iodine was forged in moments in the thermonuclear fury of a supernova explosion, whereas most barium and zirconium was cooked over thousands of millennia inside red giant stars. The paper also showed why the most abundant elements in the universe, next to hydrogen and helium, are oxygen, carbon, neon, nitrogen, magnesium, silicon and iron. They are the seven elements which have been assembled in the greatest quantities by stellar nuclear reactions.

For explaining how the vast majority of the elements in nature were made, 'B-squared-FH' is justly regarded as one of the greatest achievements of twentieth-century physics.* However, it had one major failing. It could not explain the origin of all the elements in nature. More specifically, it could not explain the origin of the very lightest elements.

Try as they might, Hoyle and his colleagues could not think of a process that could create significant quantities of either deuterium, the heavy form of hydrogen, or lithium, beryllium and boron, the third, fourth and fifth lightest elements. All were extremely heat-sensitive. Far from creating the light elements, nuclear reactions in stars actually destroyed them. In the sun, for instance, any deuterium was automatically transformed into helium-3 in the second step of the sunlight-generating proton–proton cycle.

To explain the origin of the very lightest elements, Hoyle and his colleagues were forced to propose the existence of a mysterious eighth nuclear process. They even invented a name for it. They called it the x-process. The x-process, when it was finally discovered, would account for the existence of the fragile light elements. However, there was one light element which was anything but fragile yet whose existence was nevertheless a deep mystery. The element in question was helium.

On the face of it there appeared to be no difficulty in accounting for the existence of the second lightest element in nature. After all, the majority of stars, including the sun, were ceaselessly converting hydrogen into helium. However, as Hoyle had begun to realise as early as 1955, there was in fact far too much helium in the universe. In fact,

---

* Many years later, in what most scientists consider one of the most disgraceful miscarriages of justice in the history of science, Fowler would be given the Nobel prize and Hoyle ignored. Hoyle himself believes that he lost the prize because he criticised the Nobel committee when it awarded the Nobel prize for pulsars not to their discoverer, the graduate student Jocelyn Bell, but to her supervisor.

there was at least ten times as much as the stars could possibly have made.

Where did all the extra helium come from? The answer would come from two young radio astronomers working for the Bell Telephone company in New Jersey. Their names were Arno Penzias and Robert Wilson, and their discovery would earn them both the Nobel prize for Physics.

# 15

# A Tale of Two Sites

HOW WE LEARNT THAT ATOMS WERE MADE IN
TWO PLACES – INSIDE STARS AND IN THE
FIREBALL OF THE BIG BANG.

And God said, Let there be light: and there was light.

Genesis 1:3

Arno Penzias and Robert Wilson had been lured to Bell Labs by a
very unusual radio antenna which they hoped to use for doing
astronomy.* The antenna, which stood on a low wooded knoll at
Holmdel in northern New Jersey, had been built in 1960 to bounce
radio signals off Echo 1, a 100-foot-diameter silver ball which had been
suspended in space high above the earth.

Echo 1 was the grandfather of all modern communications satellites.
It was quickly superseded by Telstar, which, rather than passively
reflecting radio signals, boosted them before beaming them back down
to earth. It was while the Telstar project was under way that Bell Labs
decided to hire a couple of radio astronomers, experienced in picking up
weak signals like those from satellites.

Penzias arrived from New York's Columbia University in 1962.
Wilson arrived from Caltech in the spring of 1963. Since they were the
only two radio astronomers in the company, they immediately teamed
up. The Telstar project had prevented Penzias from doing any radio
astronomy since his arrival at Bell Labs. Things changed, however,

---

* An antenna is any device that collects radio waves. A television aerial is an antenna; it
collects radio waves from a TV transmitter. Other examples are the giant bowl-shaped
dishes astronomers use to pick up faint cosmic radio signals.

shortly after Wilson came on the scene. The satellite engineers completed their work and departed, leaving the Holmdel antenna entirely free.

The antenna on Crawford Hill was a very unusual shape. If it resembled anything at all, it was a giant ice-cream cone lying on its side. In the body of the cone, just beneath where the giant scoop of ice-cream should have gone, was a 20-foot square opening. This acted as the 'bucket' which collected the radio waves from the sky. They were then funnelled down the interior of the cone to a wooden shed strapped to the tapered end. The shed contained the 'receiver' which actually registered the radio waves.

The reason the antenna had such a peculiar shape was to keep out unwanted radio waves. Such waves are generated by all manner of bodies in the environment – buildings, trees, even the ground – and can easily swamp a weak radio signal like the one from a tiny satellite. But radio waves have difficulty bending round sharp corners, a fact exploited by the designers of the Holmdel antenna. They had made the edges of the 20-foot aperture as sharp as possible so that when it was pointing at the sky unwanted radio waves from the environment found it extremely difficult to bend their way in.

It was the ability of the Holmdel antenna to pick up a very weak radio signal, in the midst of a host of other confusing signals, which was the big attraction for Penzias and Wilson. Their plan was to try and detect a faint halo of hydrogen gas which they suspected might be surrounding the starry 'disc' of the Milky Way. If it existed, such a halo would glow very weakly with radio waves.

But the Holmdel antenna, though good at excluding unwanted radio waves, was not perfect. It was unavoidable that some radio waves from the environment would find their way in, and these could be stronger than the radio waves from any faint galactic halo. Before embarking on their observations, Penzias and Wilson therefore decided to identify all sources of unwanted radio waves, and measure how big they were.

The obvious wavelength at which to search for the Milky Way's radio glow was 21 centimetres, since this is the wavelength at which hydrogen atoms floating in the cold of interstellar space broadcast like tiny radio stations. However, the Telstar engineers had left behind a radio receiver tuned to a wavelength of 7.35 centimetres. It therefore seemed sensible to Penzias and Wilson to try and understand the behaviour of the antenna at this wavelength, before going to the trouble of building a receiver sensitive to 21 centimetres.

At a wavelength of 7.35 centimetres, the cold hydrogen gas in the Milky Way's halo should not be giving out any radio waves at all, which provided a simple and elegant way for Penzias and Wilson to test their equipment. If they pointed the 20-foot opening of the antenna at the sky and accounted for all unwanted radio waves, the signal left over should be precisely zero.

When Penzias and Wilson carried out this test in June 1964, it was immediately clear that something was seriously wrong. Even after accounting for every conceivable source of unwanted radio waves, the Holmdel antenna still seemed to be registering a signal. It was small but absolutely unmistakable. It was exactly what would be generated by a body just 3.5 degrees above absolute zero.[*]

At first, Penzias and Wilson thought they might be picking up a man-made signal from the urban environment of northern New Jersey. The best place to do radio astronomy is an isolated valley, shielded from all radio interference. But the Holmdel antenna had been put on top of a hill so it could 'see' a satellite anywhere in the sky.

If the spurious radio signal was man-made, the obvious source was New York City, 30 miles to the north. However, the signal did not get noticeably bigger when Penzias and Wilson pointed their antenna at the city. In fact, it remained unaltered wherever in the sky the antenna was pointed. This ruled out both the sun and the Milky Way as the source since each occupied only a small portion of the sky.

The only other possibility was that there was a fault in the antenna itself. As soon as Penzias and Wilson started looking, their gaze fell on a pair of pigeons roosting deep inside the giant ice-cream cone antenna. The pigeons had coated the interior with pigeon droppings. Was it possible, the astronomers wondered, that the droppings were glowing softly with radio waves and so causing the spurious signal? There was only one way to find out. The two astronomers trapped the birds and despatched them in the company mail to a faraway Bell Labs site in New Jersey. They then climbed into the gloomy interior of the antenna with stiff brooms and swept away the droppings.

Confident that they would at last be able to get on with their observations, Penzias and Wilson then pointed the 20-foot antenna at the sky. To their dismay, they found that the annoying signal was still there. The Holmdel antenna was continuing to register an anomalous temperature of 3.5 degrees above absolute zero.

By now the excess signal had persisted, unvarying, for the best part of a year. This meant that it could not come from a source in the solar system, since any such source would necessarily have moved around

---

[*] Absolute zero is the lowest temperature attainable. When an object is cooled, its atoms move more and more sluggishly. At absolute zero (which on the Celsius scale is equal to −273.15°C), the atoms are moving as slowly as possible.

the sky as the earth orbited the sun. Penzias and Wilson had now done everything they could think of to locate the source of the spurious signal. However, just when they were ready to abandon the project altogether, Penzias happened to make a phone call to a radio astronomer friend called Bernie Burke.

Penzias wasn't calling about the peculiar signal being picked up by the Holmdel antenna but about another matter altogether. It was inevitable, however, that once the two men got chatting about their work the subject would get mentioned. When it did, Burke said that he had heard a team of physicists from Princeton University was planning to search for a faint radio hiss from space. Could the signal Penzias was describing be what they were after?

The possibility that the anomalous signal was real had been dismissed by Penzias and Wilson. After all, the fact that it was coming from every direction in the sky could only mean that the entire universe was glowing with radio waves. As it happened, this was precisely the conclusion that had been reached by the leader of the Princeton team, Robert Dicke. If the universe had once gone through an extremely hot and dense phase, Dicke realised, the whole of space would indeed be aglow with feeble radio waves.

Penzias, alerted by Burke, phoned Dicke. The result was a visit from the Princeton group a few days later. They confirmed very quickly that what Penzias and Wilson had found was the ubiquitous cosmic glow they were after. However, unknown either to Dicke or to Penzias and Wilson, the glow had actually been predicted 17 years earlier – by two students of George Gamow.

## AFTERGLOW OF CREATION

Gamow had, of course, been the first to recognise that the universe had begun in an ultra-dense, ultra-hot state. Unaccountably, however, the father of the hot big bang had overlooked one important consequence of the idea. It concerned the fierce light of the fireball of the big bang. If Gamow's students Ralph Alpher and Robert Herman weren't mistaken, it should still be around today.

It would have faded, greatly cooled by the expansion of the universe in the aeons that had elapsed since the big bang. Rather than corresponding to a temperature of a few thousand degrees, it would now correspond to a temperature of only a few degrees above absolute zero. Instead of appearing at visible wavelengths, it would appear in the radio region of the spectrum.

Alpher and Herman made their prediction of the left-over radiation from the big bang in 1948, but it was largely ignored, chiefly because the theory on which it was based had been discredited. Most of the universe's heavy elements were not created at the beginning of time, as Gamow had hoped, but had originated subsequently in the hearts of stars. When Gamow's theory of the origin of the elements was abandoned, so too was the idea of a hot big bang. It was a classic case of the baby going out with the bath water.

Now, 17 years after Alpher and Herman's prediction, Penzias and Wilson had stumbled on the left-over radiation of the big bang. The 'afterglow of creation' was the oldest fossil in existence. It had been travelling across space for 15 billion years and carried with it priceless information about what the universe had been like shortly after its birth. It was strong evidence that there really had been a blisteringly hot explosion at the beginning of time.

Scattered jigsaw pieces began to come together. Hoyle and his colleagues, despite their success in explaining the origin of the majority of the heavy elements, had never been able to explain how the universe had ended up with around 25 per cent of its mass in the form of the second lightest atom, helium. If the helium had been cooked from hydrogen inside stars, then the inevitable by-product was starlight. But if you summed up all the starlight streaming through the universe, it fell far short of what would be expected if stars had transformed 25 per cent of the mass in the universe into helium. At most, the stars could have processed a mere few per cent.

By the early 1960s, Hoyle and his student Roger Tayler had concluded that the helium puzzle could have only one solution. At some time in the past, most, if not all, of the matter in the universe must have passed through a multi-billion-degree furnace which had transmuted tremendous quantities of hydrogen into helium. Hoyle and Tayler's candidate for the furnace was an exploding 'superstar' weighing millions of times more than an ordinary star. If such a star started out hotter than 10 billion degrees, then in the first 100 seconds of its explosion, before the material became too spread out and cool for element-building, roughly 25 per cent of the star's material would be converted into helium.

The problem was that there was no evidence for superstars.[*] And they had to exist in extraordinary numbers if they were to have

---

[*] For a while Hoyle harboured the hope that the newly discovered 'quasars' might instead fit the bill. However, they turned out not to be stars but the prodigiously luminous cores of newborn galaxies at the edge of the observable universe.

processed so much of the material in the universe into helium. Hoyle had plumped for them simply because he was wedded to the idea that the universe had not been born in a one-off big bang but had existed forever. This caused him to ignore a simpler and more obvious solution – that the hot big bang had been responsible.

As early as the 1940s, Gamow had shown how element-building nuclear reactions in an expanding and cooling fireball would have efficiently converted hydrogen into helium. His calculations were primitive. However, more sophisticated ones, carried out in the 1960s by Hoyle, Willy Fowler and Robert Wagoner, bore out his general conclusions.* A second or so after the moment of creation the universe would have been an angry whirl of electrons, positrons, neutrinos and photons all flying about at an unimaginably high temperature. In the midst of the maelstrom would have been a sprinkling of much heavier particles – the protons and neutrons that would eventually combine to form the nuclei of atoms.

At first, even the labels 'proton' and 'neutron' would have meant very little; frequent collisions with lighter particles would trigger a host of nuclear reactions perpetually transforming protons into neutrons, and vice versa. However, a neutron is very slightly heavier than a proton, which means that it takes marginally more energy to make. In the beginning, when every particle was flying about so fast it carried more than enough energy to convert a proton into a neutron, this extra energy-cost was of no consequence and protons and neutrons were equally common. However, as the fireball of the big bang continued to expand and cool, the amount of energy carried by the particles fell until a time came when the greater energy-cost of making neutrons began to make itself felt. The less costly protons started to win out at the expense of the neutrons.

This then is the key to understanding why the universe contains the amount of helium it does. For when the cosmic expansion had cooled the universe to the point where there was no longer enough energy to interchange protons and neutrons, there were more protons around than neutrons. In fact, calculations show that there were between 6 and 7 times as many protons as neutrons.

Since there were roughly 7 protons for every neutron, there were 14 protons for every 2 neutrons. This is important because it takes 2 neutrons, in addition to 2 protons, to make a helium nucleus. That leaves a surplus of 12 protons. So the result of incorporating all the

---

* Hoyle, to his great credit, has never been above contributing to theories with which he has disagreed, such as the big bang theory, which he actually named.

universe's neutrons into helium is a mix containing 1 helium nucleus for every 12 hydrogen nuclei.

Now a helium nucleus weighs 4 times as much as a hydrogen nucleus. The weight of a helium nucleus plus 12 hydrogen nuclei is 16 times that of a hydrogen nucleus. So the fraction of the universe's mass which the furnace at the beginning of time turned into helium is 4 divided by 16, or 25 per cent. And 25 per cent is roughly the amount of helium we observe in the sun and stars.*

But the mere fact that a hot big bang could have produced helium in the right proportions, though convincing, was not necessarily proof that it actually did. That proof would come from examining the manufacture of other light elements, in particular deuterium and lithium.

## A TALE OF TWO SITES

The origin of deuterium remained a puzzle because stars, rather than making it, actually destroy it. The formation of deuterium from hydrogen is the first step in the proton–proton cycle by which stars like the sun generate the light and heat which give us life. But, though the first step in the proton–proton cycle makes deuterium, the second step unmakes it. For, in the hydrogen-rich environment of a star, deuterium nuclei are constantly being struck by stray protons and promptly transformed into nuclei of helium-3, the light form of helium.

Consequently, there should be virtually no deuterium in the universe. However, this is far from the case. Astronomers observe its characteristic spectral fingerprint in gas clouds floating between the stars or between the galaxies. There are in fact a few deuterium nuclei for every 100,000 hydrogen nuclei.† Since stars destroy deuterium, the mystery is where does it come from? Here, once again, the answer lies with the hot big bang.

There is a tendency for every element to be processed into helium, since helium is the most tightly bound of all the light nuclei and, to reach it, nuclear fusion reactions are effectively running 'downhill'. If this did happen, deuterium would have been wiped out along the way, as surely as it is inside stars. However, the fireball of the big bang expanded and cooled very quickly, ensuring that nuclear reactions were snuffed out just before they could run to completion.

---

* In fact, because there were between six and seven protons to every neutron at the time of helium-building in the big bang, the figure is closer to 23 per cent.

† This might not seem much, but of course hydrogen is much more common than anything else. Deuterium is more abundant than most of the heavy elements in nature.

The amount of deuterium formed is very sensitive indeed to the conditions prevailing at the time of its formation. If there are a lot of protons or neutrons around, a deuterium nucleus will very quickly snap one up and transform itself into another kind of nucleus.

We can infer the conditions prevailing in the furnace at the beginning of time by imagining the expansion of the universe in reverse. When theorists do this and calculate the amount of deuterium that should have been produced, they find it agrees very closely with what astronomers observe when they look out in the universe today.

The other light element whose origin posed a great mystery is lithium, the third lightest after hydrogen and helium. Lithium-7 is destroyed at temperatures of a million or so degrees and even the coolest stars are considerably hotter than this. So where does the universe's lithium come from? Once again, the big bang comes to the rescue. A few of the neutrons present would have stuck to helium to build heavier elements, making a few lithium atoms for every 10 billion atoms. This is very similar to the amount of lithium astronomers see in old stars whose surfaces have been too cool for fusion to take place and lithium to be consumed.

The prediction of the abundances of the light elements is one of the great triumphs of the big bang theory. It leaves astronomers in very little doubt that two main sites were responsible for the creation of atoms – the hot big bang for the very lightest atoms, and stars for all the rest.

But, as is the case with all sweeping statements in science, this is not entirely true. A few elements remain whose abundances cannot be explained simply by nuclear reactions in the big bang or inside stars. For instance, beryllium and boron, the fourth and fifth lightest elements respectively, do not fit into this picture. They are produced in stars in minimal quantities and destroyed by temperatures lower than those involved in the fusion of hydrogen to helium. Yet they do exist in the universe.

The clue to the origin of beryllium and boron comes from cosmic rays, super-fast nuclei that continually bombard the earth from space. Cosmic rays, probably the debris of violent supernovae explosions, contain a higher proportion of beryllium and boron than material either from the earth or the sun. The generally accepted explanation is that cosmic rays started off with the same element composition as any other matter, but that collisions between cosmic ray nuclei and protons floating in interstellar space caused some nuclei to break up *en route* to the earth. In this picture, beryllium and boron are nothing but nuclear shrapnel.

Mysteries still remain concerning the light elements. However, no scientist seriously doubts that the broad picture of element build-up in the big bang is fundamentally correct. The light elements are fossilised relics from the early universe, and their abundances are directly connected to the extraordinary conditions in the first few minutes of creation.

As far as we human beings are concerned, however, it is the heavy elements, not the light elements, which really matter. They, after all, make life possible. And the key to the production of heavy elements are supernovae.

# 16

## *The Engines of Creation*

HOW A DEATH IN THE NEIGHBOURHOOD
HIGHLIGHTED THE WAY IN WHICH HEAVY
ELEMENTS ARE VOMITED INTO SPACE
AND ENRICH THE GAS DRIFTING
BETWEEN THE STARS.

All humans are brothers. We came from the same supernova.

Allan Sandage

What is precious is never to forget.
The essential delight of blood drawn from ageless springs
Breaking through rocks in worlds before our earth.

Stephen Spender

A star goes supernova in our galaxy about once every 20 years. We know this because in other galaxies – at least in those which are similar in size and shape to the Milky Way – this is how often stars are observed to blow themselves apart.* Paradoxically, supernovae in our own galaxy almost always go unnoticed. The reason is dust. It chokes the star-lanes, soaking up visible light like a sponge. Because of interstellar dust, it is extremely difficult to see ordinary stars which are more than a few thousand light years away, a mere tenth of the distance from the sun to the centre of the Milky Way.†

---

* There are about 1000 billion galaxies within the observable universe. If, on average, each of these galaxies is host to a supernova explosion once every 20 years, then the universe experiences 1000 supernovae every second.

† Fortunately, radio waves and infrared light are able to penetrate interstellar dust. With telescopes sensitive to these types of light, astronomers have been able to see to the very heart of our galaxy.

On rare occasions, however, a supernova undergoes a detonation close enough to the solar system to be visible with the naked eye. Over the course of recorded history there have been many such events. In 1054, for instance, Chinese and Japanese astronomers observed a 'new star' in the constellation of Taurus. It was so bright it was visible in broad daylight for more than three weeks. The cooling remnant of the supernova — a shell of expanding matter riddled with filaments of incandescent gas — is still visible in the night sky as the Crab Nebula.

More recently, supernovae were observed in both 1572 and 1604. The first, known as 'Tycho's star' after Tycho Brahe, the Danish astronomer who recorded the event, was visible to the naked eye for 18 months. The second, christened 'Kepler's star' after Johannes Kepler, the man who discovered that the planets follow elliptical not circular paths around the sun, burned in the sky for just over a year. The supernova of 1604 was badly timed. Had it erupted on the scene a mere five years later it might have been observed by Galileo through the very first astronomical telescope. Missing this opportunity might not have mattered much if another supernova had come along quickly. However, supernovae have a habit of arriving in bunches, with long gaps in between. Little more than three decades had separated the stellar suicides of 1572 and 1604, yet there was no hint of a successor in the next 300 years. Fortunately, however, the situation recently changed.

## A DEATH IN THE NEIGHBOURHOOD

The first hint of the supernova came from a Canadian astronomer working at the Las Campanas observatory high in the Andes of Chile. In the early hours of 24 February 1987, Ian Shelton developed a photograph taken with one of the observatory's telescopes. It showed one of the Milky Way's two satellite galaxies, the Large Magellanic Cloud.

When Shelton looked at the photograph, he was surprised to see a brilliant star where no star had been before. Among the other astronomers he told was a Chilean, Oscar Duhalde. An hour earlier he had looked up at the sky and seen a new star in the Large Magellanic Cloud. Duhalde, it turned out, was the first person in 383 years to see a supernova with the naked eye.

It was christened Supernova 1987A. Since the explosion had occurred in a region of the sky which had been much photographed over the years, it was possible to go back to old plates and identify the exact star which had detonated. The cataclysm had occurred at the edge of a giant cloud of glowing gas known as the Tarantula Nebula. The gas was glowing because it was being heated by the fierce light of a clutch of hot stars which had only recently been spawned by the nebula. One of the young stars turned out to be the progenitor of Supernova 1987A.

It was called Sanduleak −69° 202 and it was an extraordinarily luminous blue-hot star.* In fact, it pumped out so much light − 100,000 times more than the sun − that it could be seen on the very first long-exposure photograph ever taken of the Large Magellanic Cloud at the end of the nineteenth century. Once astronomers had identified the precursor star, its history could be pieced together. Sanduleak −69° 202 was a huge star, 20 times more massive than the sun, and so had raced through its life at enormous speed.

The star had congealed out of the gas of the Tarantula Nebula a mere 20 million years ago, a blink of the eye on the cosmic scale. By comparison, the sun is more than 200 times this age. When the temperature in the core of Sanduleak −69° 202 had risen to 10 million degrees, hydrogen nuclei began to stick together to make helium via the carbon-nitrogen cycle. The fusion of hydrogen liberates so much more nuclear binding energy than any other conceivable nuclear reaction that it can keep a star shining for 95 per cent of its lifetime. However, Sanduleak −69° 202 had a hotter core than the sun on account of its high mass, causing nuclear reactions to proceed far more quickly. By a million years ago, all the hydrogen in the star had been turned into helium, and the star faced a crisis.

The star's helium core shrank and the temperature rose. When it reached 50 million degrees, nuclear reactions began to turn helium into carbon via the triple-alpha process. At the same time, the outer layers of the star swelled, turning the star into a red super giant hundreds of times bigger than the sun. In fact, the star became so big that its gravity could not hold on to the outermost regions and the gas in these regions sailed off into space as a 'stellar wind'. The wind may have siphoned off a quarter of the mass of Sanduleak −69° 202, causing it to shrink back until it was a blue giant only 50 times bigger than the sun.

All the while the star's core was approaching another crisis. The inner portion had now turned to carbon. The star's central parts reacted by shrinking again; the temperature of the centre rose, and carbon began to 'turn' into neon. The carbon 'fuel' lasted only a thousand years before the centre shrank again and heated, and the neon turned to silicon.

* The star is named after Nicholas Sanduleak of the Warner and Swasey Observatory in Cleveland, Ohio, who catalogued it in the 1960s. Sanduleak included the star in his catalogue as number 202 in the band at 69° south of the celestial equator − hence its catalogue designation, Sanduleak −69° 202.

Now the star was in a desperate state. The amount of nuclear energy that can be provided by nuclear fusion decreases as heavier and heavier nuclei 'burn'. The nucleus beyond silicon is iron, and iron nuclei are the most stable and tightly bound of all. If iron is fused with other nuclei, no more nuclear binding energy is liberated. In fact, such a reaction will suck heat from the star.

Within a year, the star had 'burnt' all its neon to silicon. The central temperature rose to 400 million degrees and the last available nuclear reaction began. At the centre silicon fused to make iron, a reaction which was completed in only a couple of days. The star now lacked a source of energy to support the compacted core against the tremendous pull of its own gravity.

The star's interior at this point consisted of successive shells of matter that had undergone different amounts of nuclear 'burning'. At the centre was a core of iron with a mass about 50 per cent greater than that of the entire sun. Around it were shells consisting mainly of silicon, neon, carbon and helium, with a total mass equivalent to six suns. On the outside was a shell consisting mainly of pristine hydrogen, unchanged since the birth of the star. With the termination of the nuclear reactions in the iron core, the star's central regions shrank catastrophically under their own gravity.

Nobody is entirely clear how the collapse of the central region of a star turns into the explosive expansion of its outer regions. A strong possibility, however, is that this is accomplished by neutrinos, the most ghostly and insubstantial subatomic particle in nature's arsenal.

Neutrinos are produced in copious quantities when electrons and protons are jammed together to make neutrons. Free electrons are always around in the star. The protons come from nuclei banged together so violently they shatter, undoing a star's laborious atom-building in a split-second. When the protons and electrons in the centre of the star combine to make neutrons, the result is a 'neutron core', a super-dense relic about 20 kilometres across but weighing about as much as the sun, and a flood of neutrinos.

Normally, neutrinos pass through matter as easily as photons of light pass through transparent glass. But the super-dense collapsed matter in the centre of the star impedes their progress. Unable to slide through, the neutrinos drive all before them like the blast wave from a bomb. As a consequence, the outer layers of the star surge out into space.

Experiments on earth picked up a total of 19 neutrinos – the first ever detected from a celestial body beyond the sun. For the experimenters in Japan and America it was an enormous triumph.

If neutrinos do indeed blow stars apart in supernovae – and this is by no means proven – then here we have one of nature's greatest ironies. For the most insubstantial particles in the universe are behind the most catastrophic of cosmic events.*

# THE ENGINES OF CREATION

Supernova 1987A provided an opportunity for astronomers to see at first hand the creation of elements and their scattering to the winds of space. Evidence of the actual creation of elements in the inferno of the explosion came from the way in which the light from the supernova varied in the months immediately after the explosion.

At its peak of brightness, Supernova 1987A could have easily outshone 250 million suns. So brilliant was the stellar funeral pyre that people in the southern hemisphere could see it easily even when the glare from the moon or from street lamps completely obliterated the diffuse light of the Large Magellanic Cloud.

The puzzle was that the supernova stayed so bright for so long. Its peak of brightness had come on 20 May, three months after Sanduleak −69° 202 had detonated. Contrary to expectations, the expanding fireball had not cooled and faded rapidly. By the end of May, there should have been nothing more than an insignificant glow. It was clear that something was keeping the fireball hot and shining brilliantly.

The explanation came from two American theorists, David Arnett and Stan Woosley. Independently of each other, they suggested that what was keeping the fireball hot was radioactivity from nickel-56. According to Arnett and Woosley, the isotope was created when the blast wave ripping its way out from the core of the star tore into the onion-skin layer that contained oxygen. So violently did this compress and heat the layer that it triggered an outburst of nuclear reactions. In a flash, fusion reactions build up oxygen nuclei into nickel-56.

According to Arnett's calculations, the mass of nickel-56 created in Supernova 1987A was equivalent to about 7 per cent of the mass of the sun. Nickel-56 quickly decays into cobalt-56, which in turn decays into iron-56 over a period of a few months. More heat is actually liberated in this decay process than in the initial supernova explosion. Shortly after the detonation, nickel-56 would be hidden deep within the star so the extra heat would not show up. But, as the star's outer layers expanded and thinned, it would be possible to see deeper and deeper into the star

---

* In fact, the neutrinos carried off 99 per cent of the supernova's energy. The expanding fireball that shone in our skies was a mere relic of the intense neutrino burst.

until, eventually, the layers of gas being heated by the radioactivity would be exposed.

Confirmation of this picture was not long in coming. Having peaked in brightness, the supernova began to fade at exactly the rate that cobalt-56 decays. Then, towards the end of 1987, instruments on satellites and platforms lifted to the edge of space by high-altitude balloons picked up gamma rays with the characteristic energy signature of cobalt-56. Even more dramatic were observations of the optical spectral line of cobalt. Over the next couple of years, astronomers at the Anglo-Australian telescope in Australia saw the line gradually fade in the supernova's spectrum. The element was turning into iron before their eyes.[*]

## THE COSMIC STOCKPOT

Supernovae are the engines of creation. Not only do they give birth to new elements, but they scatter those elements to the currents of space along with many of the elements forged in the normal course of a star's life. This return of elements to space is one of their most important roles. Whereas red giants and other mildly unstable stars expel mostly medium-weight elements such as carbon and oxygen, supernovae alone eject significant quantities of the very heaviest elements.

When Sanduleak $-69°$ 202 went supernova, its surface layers exploded into space at something like 10,000 kilometres per second. At first, the debris from the explosion was unstoppable. It swept all before it, carving out an ever-growing cavity in the surrounding interstellar medium. However, the gas of the interstellar medium is not without some resistance. In the same way that snow impedes the progress of a snow plough, so the interstellar medium impedes the progress of anything travelling through it. The effect it has is tiny, as the gas between the stars is extremely thin; there is barely one atom rattling about in every sugar-cube-sized volume of space.[†] However, with every passing year, more interstellar gas piles up in front of the expanding debris of Supernova 1987A. Pushing the extra weight becomes ever harder. Eventually, perhaps in a few thousand years' time,

---

[*] Supernova 1987A was a so-called Type II supernova in which an ultra-massive star explodes. In a Type I supernova, which involves the detonation of a highly compacted star, 20 times more matter can be converted into radioactive nickel-56. In consequence, Type I supernovae are correspondingly brighter.

[†] This is far more rarefied than even the best vacuum scientists have succeeded in creating on earth.

the weight will be so great that the supernova debris will grind to a complete halt.

Long before this happens, however, the expanding shell of supernova detritus will begin to break up. No supernova explosion is completely uniform, and neither is the interstellar medium through which the ejecta travels. These factors cause some material from the explosion to lag behind other material from the explosion. Where once there was a neat spherical shock front spreading outwards like a ripple on a pond, fingers of matter will begin to extend outwards into space. The result of this can be seen in the remnant of the supernova of 1054, whose dramatic crab-like appearance explains why it is known as the Crab Nebula.

The effect of the slow-down and break-up of the debris from Supernova 1987A will be to mix thoroughly the heavy elements cooked inside Sanduleak −69° 202 with the gas of the interstellar medium.

The American astronomer Laurence Marschall compares the interstellar medium to a stockpot which a farmer's family keeps simmering on the stove.* Into the pot go spices, herbs, vegetables from the garden and occasional left-overs of game and poultry. At intervals the members of the family dip into the pot for meals and when the pot runs low each replenishes it with a favourite herb or piece of food. Consequently, on any one day, the balance of ingredients in the stock reflects the history of its making. If the garden produced a lot of carrots, for instance, or the farmer killed a goose recently, the stockpot shows it.

Marschall points out that the pattern of element abundances in our galaxy is very much like the mix of ingredients in the stockpot. It too reflects the history of its making. Just as different members of the family contribute to the overall balance of the stock, so different celestial bodies contribute to the overall balance of the interstellar medium and the relative abundances of different elements reflects this. Supernovae do their bit, as do red giants and other types of stars. Each forges its own characteristic assortment of elements and mixes a portion of them back into the common pool.


## THE ASH OF LONG-DEAD STARS

But stars do more than simply add to the common pool. They take from it as well. For the gas of the interstellar medium is the raw material out of which new stars are made.

And here it appears that supernovae have another critical role to play.

---

* Cf. Marschall, *The Supernova Story* (Princeton University Press), p. 211.

For interstellar gas is apt to float around in cold, dark clots, rarely changing or reacting in any way.

Supernovae, however, can be catalysts for change. When the high-speed debris from a supernova explosion slams into an interstellar cloud, the tremendous impact compresses the gas. This may be all that is required to change the cloud's dormant state, and set globs of it shrinking under the force of their own gravity. When these become dense and hot enough to ignite hydrogen-burning nuclear reactions, the result is a clutch of newborn stars.

If this scenario is correct, as many astronomers suspect it is, then suicidal stars such as Sanduleak −69° 202 have a key role to play in kick-starting the whole process of star formation. In more ways than one, supernovae are the engines of creation.

At the time of its birth, a star's composition reflects the current composition of the interstellar medium, just as a meal from Marschall's stockpot reflects the current composition of the pot. Of course, the composition of the interstellar medium may differ from location to location within our galaxy, depending on, for example, how crowded together the stars are, and how frequently supernovae are going off. But since stars from one end of the galaxy to the other are continually vomiting nuclear ash into space, the interstellar medium at every location gets richer and richer in heavy elements as time goes by.

The obvious implication of this is that in the past the galaxy was poorer in heavy elements than it is today. This is confirmed by examining the light of the very oldest stars. Those which formed shortly after the birth of the galaxy – some 10 billion years or so ago – have a very meagre complement of heavy elements. In fact, some very ancient stars have less than a thousandth of the concentration of iron of the sun.

Since the birth of these ancient stars, the concentration of heavy elements in the interstellar medium has climbed remorselessly. If it were somehow possible to view a movie of the history of our galaxy, and fast-forward it from Day One to the present, we would see supernovae going off like stuttering firecrackers. Something like a billion stars have detonated since the Milky Way was born.*

None of these stellar deaths was in vain. For each scattered heavy elements to the winds of space. And these elements were not wasted but incorporated into new stars which congealed out of interstellar clouds enriched by the supernovae debris. Countless stars have died since the birth of the galaxy and countless other stars have risen, phoenix-like, from the ashes.

---

* Most astronomers believe that the rate of supernovae was higher in the distant past than today.

Some of these stars in turn lived out their lives and died as supernovae. The cosmic cycle of death and rebirth has been repeated many times in the history of our galaxy. The sun, for instance, is often reckoned to be a third-generation star. This means that two generations of stars lived out their lives and died, in the process seeding the interstellar medium with the ash of their nuclear furnaces, long before the sun was born. Without those earlier generations of stars, the sun would not possess the rich complement of heavy elements that it does. There would be no rocky planets like Mercury, Venus, the earth and Mars, since these are predominantly made of heavy elements such as iron and nickel and aluminium. And, of course, there would be no living organisms. For without heavy elements like carbon and nitrogen, phosphorus and iodine, sodium and potassium, the complex chemistry of life would be impossible.

In order that we might live stars in their billions, tens of billions, hundreds of billions even, have died. The iron in our blood, the calcium in our bones, the oxygen that fills our lungs each time we take a breath – all were cooked in the furnaces of stars which expired long before the earth was born.

That the heavy elements in our bodies directly connect us to cataclysmic events which occurred billions of years ago in our galaxy is extraordinary enough. But this is only part of the story. For the heavy elements in our bodies may also connect us to a time so impossibly remote that at present we have only the vaguest knowledge of it. That time was before the birth of the Milky Way itself.

## THE SCENT OF ANCIENT STARS

According to the story told in many popular accounts, the Milky Way congealed out of the primordial gas of the big bang, which contained only hydrogen and helium and the merest sprinkling of other light elements forged in the first minutes of the universe. Consequently, the Milky Way's very first stars were devoid of any heavy elements. However, the evidence from examining the light of some of the galaxy's most ancient stars is that even the stars most depleted in heavy elements nevertheless contain a thousandth or a ten thousandth the concentration present in the sun. None of the stars have none.

Of course, it could be that stars with no heavy elements whatsoever are simply so rare these days that we have not yet stumbled across any. Alternatively, it could be that our scientific instruments are not capable of detecting heavy elements whose abundances are less than a ten thousandth of that of the sun. However, there is a far simpler explanation of why we have seen no stars without heavy elements: that they do not exist. In other words, the Milky Way, even at the time of its birth, contained at least some heavy elements.

The evidence for this comes from an unexpected source – quasars. Quasars lurk deep in the cores of some galaxies. They are widely believed to be giant black holes – in some cases as heavy as 10 billion suns – sucking in matter and, in the process, heating it to incandescence. A quasar is typically as luminous as 100 Milky Ways.

The prodigious light output of quasars means they can be seen at enormous distances – far further away than any normal galaxies. They are beacons at the very limits probed by our telescopes, a property which makes them extremely useful astronomical tools. For the light of a distant quasar, on its mammoth journey across space to the earth, passes through galaxies and clouds of gas which litter the space in between. Most of these objects will be too faint to be seen directly even with the most powerful of telescopes. However, each, by absorbing some portion of the quasar's light, will leave its indelible mark in the quasar's spectrum. Hydrogen in an intervening object will absorb the quasar's light at the characteristic wavelengths of hydrogen, calcium at the characteristic wavelengths of calcium, and so on. The result will be a large number of bites taken out of the quasar's light – a big one for hydrogen, since hydrogen atoms are far and away the most numerous atoms in creation, and a lot of smaller ones for everything else.

When astronomers began to examine these fingerprints in detail, they found that many of the invisible objects between us and quasars contain heavy elements. In these objects the bite from hydrogen is very big, which indicates that they contain a large amount of hydrogen and are almost certainly galaxies that are too faint to be seen directly. Since they are galaxies, the origin of the heavy elements is no mystery. They were forged in the galaxy's stars, which then sprayed them into space when they went supernova, just as in our own galaxy.

But although astronomers found that many of the objects between us and quasars contained heavy elements, they also found many of the objects appeared to contain no heavy elements. In these objects the bite from hydrogen is small, indicating that they contain little hydrogen and are almost certainly tenuous clouds of gas drifting between the galaxies rather than fully fledged galaxies. The question was: were these clouds really devoid of heavy elements or was it simply that the bites made by their heavy elements – always much less pronounced than the bite made by hydrogen – were just too tiny for astronomical instruments to spot?

The answer turns out to be the latter. In the past few years astronomers have detected the faint fingerprints of heavy elements in a number of these tenuous intergalactic gas clouds. It is still too early to say anything definite but it may be that 10 per cent, or even 100 per cent, of all such clouds contain heavy elements.

This discovery has raised further questions to do with the origin of the heavy elements. How did they get into these clouds of gas floating between the galaxies? After all, the clouds contain no stars to cook the heavy elements and cough them into space. The answer can only be that the heavy elements were forged by an earlier generation of stars which died and contaminated the clouds with its collective ash.

In the universe time is synonymous with distance, since the only way we see the universe is with light, and as light takes longer to reach us from distant objects than it does from nearby ones, we see them farther back in time. Some of the intergalactic clouds are very far away – almost as far away as the quasars whose light has revealed them, quasars which are themselves close to the edge of the observable universe. To reach us from such a distance, the light has travelled for something like 14 billion years.[*] In other words, the clouds existed when the universe was only about a billion years old. The inescapable conclusion is that the stars which polluted the intergalactic gas clouds with heavy elements lived and died before this time – in the first billion years of the universe.

What were these stars and how did they come to be burning at the dawn of time? The short answer is that nobody really knows, since no telescope in the world can see objects so distant in both time and space.[†] But though the truth lies beyond the edge of the currently observable universe, we can make some guesses.

## THE COSMIC CONNECTION

In the beginning was the big bang, and the hot, glowing matter in the big bang was spread evenly throughout space. Today, roughly 15 billion years after the big bang, matter is no longer spread evenly – it is tied up in galaxies and stars. In fact, even by the time of the most distant quasars currently observable – about a billion years after the big bang – galaxies and stars already existed in profusion.

---

[*] Or rather 'absence of light', since the only way we infer the existence of such intergalactic clouds is by the bites they take out of a quasar's light.

[†] A giant new orbiting space telescope, currently being planned by NASA, promises to reveal galaxies and stars way beyond the most distant quasars so far observed. Should the $0.5 billion funding it needs be found, the Next Generation Space Telescope could be operating by 2010.

Clearly, at some point in the billion years which followed the big bang, the universe must have made the transition from smooth and featureless to clumpy and star-spangled. So how did it happen?

It seems that the matter in the big bang was not spread perfectly evenly. The origin of this initial unevenness is still the subject of debate, although some attribute it to 'quantum fluctuations' in the first split-second of the big bang. The key point, however, is that gravity always acts to magnify any unevenness, no matter how small. It takes a clumpy gas and makes it clumpier. Regions of the cooling fireball which were slightly denser than the rest had stronger gravity than average, which enabled them to pull in matter more quickly than the rest, which boosted their gravity, which enabled them to pull in matter faster, and so on. Not long after the big bang, therefore, all across space clumps of gas of all sizes began to congeal.

At some stage, these clumps began to spawn the first stars. Nobody knows precisely when this happened, but the dramatic change this brought about is easy to imagine. 'The universe switched on like a Christmas tree,' as some astronomers have put it.

The exact size of the gas clouds in which the first stars formed is also not known. What is known, however, is that it takes time for a cloud of gas to shrink, and that big clouds take longer to shrink than small clouds. It seems likely therefore that the universe's first stars were in small clouds – far less massive than present-day galaxies like the Milky Way. These then were the stars whose deaths polluted the gas between the galaxies with heavy elements. They lived out their lives and expired before the Milky Way was born – before the birth of even the most ancient quasars we have ever observed were born. They were a pre-galactic generation of stars.

So what happened to them? As with so much of this story, at present all we can do is speculate. It seems likely, however, that the small clumps of gas in which the first stars broke out like a rash collided with each other and merged to make bigger clumps of gas. Even in today's universe we can see galaxies coalescing, and this merger process is widely believed to have played a major role in the evolution of galaxies. If indeed the small clumps of gas did merge to make bigger clumps then, as time went by, they could easily have grown into the big galaxies which exist today – galaxies like the Milky Way. In the process, the burnt-out hulks of some pre-galactic stars may have actually been incorporated in our own galaxy. As well as being a nursery for young stars, the Milky Way may be a retirement home for some of the most ancient stars in the universe.

How much more complicated this makes the history of our galaxy.

But how much more fascinating and exciting too. Just think what it would be like if we discovered that living among the teeming millions in a modern-day city like London or New York was an Egyptian pharaoh or a Babylonian priest or a Roman legionnaire.

But whether any of the pre-galactic stars survives or not is incidental to this story. The point is that the mere existence of a pre-galactic generation of stars means that the Milky Way did not form out of the pristine stuff of the big bang. Long before the birth of the Milky Way – or any galaxy for that matter – some of the primordial stuff was processed into heavy elements and coughed out into space. So, whether or not the Milky Way formed from smaller clouds of gas which merged, or from a huge cloud which shrank under its own gravity, it almost certainly had some heavy elements at its birth. And some of them could easily have ended up in the solar system when the sun formed.

So, although most of the heavy elements in our bodies were certainly formed in earlier generations of stars that lived and died in the Milky Way, some may have been formed in a pre-galactic generation of stars that existed long before the birth of our galaxy, close to the beginning of time itself.

And if heavy elements existed long before the Milky Way, perhaps in rare, isolated pockets of the universe those elements were able to congeal into rocky planets. And perhaps, against all the odds, life got going on those planets, not a mere 4 billion years ago but 15 billion years.

Perhaps the reason why we have discovered no trace of ET, despite all our searching, is not that we are the first intelligence in the cosmos, as some have maintained, but that we are the last, born in the twilight of the universe. Not only did stars burn brighter before the galaxies but minds burned brighter too. Once the heavens were abuzz with celestial commerce, but that commerce fizzled out long ago. Life belonged to the blazing dawn of the universe.

This is speculation piled on speculation, of course. But, if nothing else, it illustrates one thing. Although the quest for the origin of the atoms in our bodies has brought us a long way in understanding our place in the cosmos, there is no doubt that we still have a long, long way to go.

# Epilogue

## THE ATOMS OF CURIOSITY

Out of the cradle
onto dry land
here it is
standing:
atoms with consciousness;
matter with curiosity.

Richard Feynman

We've come a long way...
First, we guessed that everything was made of atoms – tiny, indestructible grains of matter which came in only a handful of different types and whose myriad combinations gave rise to all manner of different substances. This insight, coming more than 2000 years before the proof of the existence of atoms, lifted a corner of the veil that shrouded the world from our senses, exposing the simplicity that lay beneath. A rose or a bar of gold or a human being were all simply patterns of atoms. The complexity of the world was an illusion, a reflection merely of the endless permutations of a handful of basic building blocks. Everything was in the combinations.

Much later we discovered that atoms were not the smallest things. They were composed of even tinier building blocks and the difference between, say, a carbon atom and an oxygen atom and a gold atom was simply the way in which those building blocks were put together. The idea arose that atoms had been made, build up one subatomic brick at a time from the lightest atom, hydrogen. Atoms, instead of being placed in the universe fully formed, had come in kit form.

Remarkably, the nuclear processes that had built up atoms had left their indelible fingerprint all around us and this became clear as soon as it was possible to determine which elements were common in the universe and which were rare. The abundances of each element turned out to reflect the properties of its nuclei. There could be only one explanation. Nuclear processes must have been involved in creating those abundances. It was the first compelling evidence that all elements really had been made. The hunt was now on to find where this had happened.

Eventually it was discovered that nature used two main furnaces – the inferno of the big bang and the interiors of stars. Helium and the light elements were forged in the first 10 minutes of the universe's existence, and incorporated into the first generation of stars. Most of the elements heavier than helium were gradually assembled in the central furnaces of stars and then ejected into space, principally by red giants which gently puffed off their outer atmospheres, and by supernovae which catapulted stellar debris throughout the galaxy.

We discovered that the atoms that go to make up our bodies have a rich and complex history. They were manufactured by long-dead stars which in their death throes vomited them into space. They floated out between the stars, where they enriched the gas and dust already there, adding to the 'cosmic stockpot'. New stars were born out of the material of this stockpot and the deaths of these stars in turn coughed the atoms they had forged into interstellar space.

Nobody knows for sure how many times this cycle of cosmic death and rebirth was repeated before a cold cloud of gas and dust on the outskirts of the Milky Way began to slowly shrink under its own gravity 5 billion years ago. The cloud had hung in space, inky black against the background stars, for tens of millions, perhaps hundreds of millions, of years. What set it in motion is not certain. Perhaps it was the impact of debris from a nearby supernova. Once it began shrinking, however, there was no stopping it. In time, the heavy elements in the cloud became incorporated into the newborn sun and its entourage of planets. They became incorporated into the earth and its rocks and ultimately into the first primitive living cells.

It is not known how life began. Perhaps it arose spontaneously in the chemicals that pooled on the earth's still-cooling surface. Perhaps, as some have maintained, it was seeded from space, carried on the back of an impacting comet. But however it got started, it was instantly subjected to natural selection, which weeded out those organisms less efficient than their peers at competing for scarce resources. Life left behind the primordial soup and, in the fight for survival, became ever more complex.

Finally, after millions upon millions of centuries, there arose creatures that thought, that speculated, that theorised about the world they inhabited and asked questions about how they had come to be in it. The atoms forged long ago in the fireball of the big bang, in countless stars across the length and breadth of the galaxy, became incorporated into human beings. They became, in short, the atoms of curiosity.

Now, why should the universe be constructed in such a way that atoms acquire the ability to be curious about themselves? That, surely, is one of the great unexplained puzzles of science.

# Glossary

*alpha decay*   ejection of a high-speed alpha particle by an unstable atomic nucleus. The result is a nucleus of an element with two fewer protons than the original.

*alpha particle*   bound state of two protons and two neutrons – essentially a helium nucleus – which rockets out of an unstable nucleus during alpha decay.

*alpha process*   build-up of heavy nuclei inside stars by the addition of alpha particles. Requires a temperature of about a billion degrees.

*anthropic principle*   idea that the universe is the way it is because if it was not we would not be here to notice it.

*Andromeda galaxy*   nearest big galaxy to our own Milky Way.

*angstrom*   unit used to measure the wavelength of light – a 10 millionth of a millimetre, which is also roughly the size of an atom.

*antiparticle*   all subatomic particles have associated with them antiparticles with opposite properties. For instance, the negatively charge electron is twinned with a positively charged antiparticle known as the positron.

*atom*   the building block of all matter. An atom consists of a nucleus orbited by a cloud of electrons. The positive charge of the nucleus is exactly balanced by the negative charge of the electrons.

*atomic mass*   total number of particles – protons plus neutrons – in an atomic nucleus. For instance, hydrogen has atomic mass 1, and magnesium atomic mass 24.

*atomic nucleus*   the tight cluster of protons and neutrons (a single proton in the case of hydrogen) at the centre of an atom. The nucleus contains more than 99.9 per cent of the mass of an atom.

*atomic number*   the number of protons (or, equivalently, electrons) in an atom.

For instance, hydrogen has atomic number 1, and magnesium has atomic number 12.

*'B-squared-FH'*   a milestone paper which detailed how most of the elements in nature had been forged inside stars. Published in 1957 by Margaret and Geoffrey Burbidge, William Fowler and Fred Hoyle, it is known by the initials of its four authors.

*beta decay*   the ejection of a high-speed beta particle by an unstable atomic nucleus. The nucleus left behind is of an element with one more proton.

*beta particle*   an electron ejected during beta decay. The electron does not exist in the nucleus beforehand but is 'created' when a neutron changes into a proton.

*big bang*   the explosion in which the universe is thought to have been born, between 10 and 15 billion years ago.

*big-bang nucleosynthesis*   the build-up of the light elements in the fireball of the big bang.

*binding energy*   see *nuclear binding energy*

*black hole*   the warped space-time left behind when a massive body such as a star has been crushed out of existence by its own gravity. Nothing, not even light, can escape the vicinity of a black hole.

*carbon-nitrogen cycle*   the series of nuclear reactions by which stars significantly more massive than the sun turn hydrogen into helium. It is a cycle because in the end it re-creates the carbon used by the nuclear reactions.

*cosmic background radiation*   the 'afterglow' of the big bang. Incredibly, it still permeates all of space 15 billion years after the event.

*cosmic rays*   high-speed atomic nuclei, mostly protons, from space. Low-energy ones come from the sun; high-energy ones probably come from supernovae. The origin of ultra-high-energy cosmic rays, particles millions of times more energetic than anything we can currently produce on earth, is one of the great unsolved puzzles of astronomy.

*deuterium*   a rare isotope of hydrogen. Deuterium contains a neutron as well as a proton in its nucleus.

*electron*   a negatively charged subatomic particle which orbits the nuclei of atoms. It is apparently a truly elementary particle, incapable of being subdivided.

*element*   a substance which cannot be reduced any further by chemical means. All atoms of a given element possess the same number of protons in their nucleus. For instance, all atoms of hydrogen have 1 proton; all atoms of magnesium have 12.

*equilibrium process*   a nuclear build-up process in which the nuclear reactions are so fast that a steady state is reached, in which every nuclear reaction is perfectly balanced by its opposite. Consequently, the mix of elements 'freezes out', i.e. does not change with time.

*expanding universe*   the fleeing of the galaxies from each other in the aftermath of the big bang.

*fusion*   see *nuclear fusion*

*galaxy*   one of the building blocks of the expanding universe. Galaxies are great islands of stars; our own island, the Milky Way, is spiral in shape and contains several hundred thousand million stars.

*gamma ray*   the highest energy form of light, generally produced when an atomic nucleus rearranges itself.

*half-life*   the time it takes half the atoms in a radioactive sample to disintegrate. After one half-life, half the atoms will be left; after two half-lives, a quarter; after three, an eighth, and so on. Half-lives can vary from a split second to many billions of years.

*heavy elements*   elements forged inside stars. Essentially, all elements except the five light elements.

*helium*   the second lightest element in nature, and the only one to have been discovered on the sun before it was discovered on the earth. Most helium was forged in the big bang. It is the second most common element after hydrogen.

*helium-burning*   the stage after hydrogen-burning, when a star combines helium into carbon and oxygen.

*hydrogen*   the lightest element. A hydrogen atom consists of a single proton orbited by a single electron. Close to 90 per cent of all atoms in the universe are hydrogen atoms.

*hydrogen-burning*   the fusion of hydrogen into helium, accompanied by the liberation of large quantities of nuclear binding energy. This is the power source of the sun and most stars.

*hydrostatic equilibrium*   the state in which the gravitational force trying to crush a star is perfectly balanced by the force of its hot gas pushing outwards.

*interstellar medium*   the gas and dust floating between the stars. In the vicinity of the sun this gas comprises about one hydrogen atom in every three cubic centimetres.

*ion*   an atom or molecule which has been stripped of one or more of its orbiting electrons, and so has a net positive electrical charge.

*isotope*   one possible form of an element. Isotopes are distinguishable by their differing masses. For instance, chlorine comes in two stable isotopes, with a mass of 35 and 37. The mass difference is due to a differing number of neutrons in their nuclei. Chlorine-35 contains 18 neutrons and chlorine-37 contains 20 neutrons. (Both contain the same number of protons – 17 – since this determines the identity of an element.)

*light, speed of*   the cosmic speed limit – 300,000 kilometres per second.

*light year*   a convenient unit for expressing the distance between stars. It is simply the distance light travels in one year, i.e. 9.46 trillion kilometres.

*light elements*   the three lightest elements – hydrogen, helium and lithium – which were forged in the big bang. Sometimes the term is extended to include the fourth and fifth lightest elements, beryllium and boron.

*luminosity*   the total amount of light pumped into space by a celestial body, such as a star.


*mass*   the most concentrated form of energy. A single gram contains the same amount of energy as 100,000 tonnes of dynamite.

*Milky Way*   our galaxy.


*nebula*   a cloud of tenuous gas in space. If young hot stars are embedded in the gas, they will cause it to glow brightly. If there are no such stars, it may still reveal itself as a black splotch which blots out the light of more distant stars.

*neutrino*   a neutral subatomic particle with a very small mass that travels very close to the speed of light. Neutrinos hardly ever interact with matter. However, when created in huge numbers they can blow a star apart, as in a supernova.

*neutron*   one of the two main building blocks of the atomic nucleus at the centre of atoms. Neutrons have essentially the same mass as protons, but carry no electrical charge. They are unstable outside of a nucleus, and disintegrate in c. 10 minutes.

*neutron star*   a super-dense ball of neutrons left over after a supernova explosion. A neutron star contains about the same amount of matter as the sun but compressed into a region only as big as Mount Everest.

*nova*   a star which explodes but whose luminosity is not as great as that of a supernova.

*nuclear energy*   the excess nuclear binding energy released when one atomic nucleus changes into another atomic nucleus.

*nuclear binding energy*   the energy possessed by an atomic nucleus by virtue of the fact that it takes energy to pull its components apart. Binding energy is released when a nuclear transformation results in a nucleus with less energy per nuclear particle. This happens in the radioactive decay of a heavy element into a lighter element and in the fusion of a light element into a heavy element. This latter process is exploited by the sun. The binding energy unleashed by the fusion of hydrogen into helium is the ultimate source of sunlight.

*nuclear fusion*   the welding together of two light nuclei to make a heavier nucleus, a process which results in the liberation of nuclear binding energy.

*nuclear reaction*   any process which converts one type of atomic nucleus into another type of atomic nucleus.

*nucleon*   an umbrella term used for protons and neutrons, the two building blocks of the atomic nucleus.

*nucleosynthesis*   the gradual build-up of heavy elements from light elements, either in the big bang – big bang nucleosynthesis – or inside stars – stellar nucleosynthesis.

*nucleus*   see *atomic nucleus*

*photon*   a particle of light.

*planet*   a small sphere-shaped body orbiting a star. A planet does not produce its own light but shines by reflecting the light of its star.

*plasma*   an electrically charged gas of ions and electrons.

*pre-galactic generation of stars*   a hypothetical generation of stars which formed shortly after the big bang and went through their lives before the Milky Way and other present-day galaxies were born.

*Population I stars*   stars found in the Milky Way's 'spiral arms'. Population I stars are young, and are relatively rich in heavy elements.

*Population II stars*   cool, red stars found in the central region of the Milky Way. Population II stars are old, and relatively poor in heavy elements.

*proton*   one of the two main building blocks of the nucleus. Protons carry a positive electrical charge, equal and opposite to that of electrons.

*proton-proton chain*   the chain of nuclear reactions by which stars of masses up to one and a half times that of the sun turn hydrogen into helium.

*quasar*   a galaxy which derives most of its energy from matter heated to millions of degrees as it swirls into a central giant black hole. Quasars can generate as much light as a hundred normal galaxies with a volume smaller than that of the solar system, making them the most powerful objects in the universe.

*quasar absorption lines*   bites taken out of the spectrum of a quasar by gas floating between the quasar and the earth. Since quasars are some of the most distant objects visible, these bites provide a means of estimating the abundances of different elements at huge distances across the cosmos.

*r-process*   build-up of heavy nuclei by adding neutrons, one at a time, to a seed nucleus. The neutrons arrive so quickly that any unstable nucleus that happens to be formed has no time to beta-decay before the arrival of the next neutron. The r-process is thought to operate in supernovae, and leads to nuclei such as iodine and gold.

*radioactive decay*   the disintegration of heavy atoms which are unstable into lighter atoms. The process is accompanied by the emission of either alpha particles, beta particles or gamma rays.

*radioactivity*   see *radioactive decay*

*red giant*   a star which had exhausted the hydrogen fuel in its centre and swollen dramatically in size. Such a star can be any colour, but red is most common.

*s-process*   build-up of heavy nuclei by adding neutrons, one at a time, to a seed nucleus. The neutrons arrive so infrequently that any unstable nucleus that happens to be formed always has time to decay before the arrival of the next neutron. The s-process operates in red giants, and forges elements such as barium and zirconium.

*solar system*   the sun and its family of planets, moons, comets and other assorted rubble.

*spallation*   the break-up of cosmic ray nuclei when they hit protons floating in interstellar space *en route* to the earth. Provides an explanation for some light nuclei such as beryllium and boron, which are regarded as nuclear shrapnel.

*spectral line*   atoms and molecules absorb and give out light at characteristic wavelengths. If they swallow more light than they emit, the result is a dark line in the spectrum of a celestial object. Conversely, if they emit more than they swallow, the result is a bright line.

*spectroscopy*   the technique of measuring the spectrum of an object.

*spectrum*   the separation of light into its constituent 'rainbow' colours.

*star*   a giant ball of gas, which replenishes the heat it loses to space by means of nuclear energy generated in its core.

*strong nuclear force*   the powerful short-range force which hold protons and neutrons together in an atomic nucleus.

*subatomic particle*   a particle smaller than an atom, such as an electron or a neutron.

*supernova*   a cataclysmic explosion of a star. A supernova may, for a short time, outshine an entire galaxy of 100 billion ordinary stars. It is thought to leave behind a highly compressed neutron star.

*Supernova 1987A*   a supernova discovered in the Large Magellanic Cloud, a satellite galaxy of our galaxy, on 24 February 1987. It was the first supernova to be seen in our neighbourhood for 387 years.

*sun*   the nearest star to our planet.

*technetium*   a radioactive element with a half-life measured in millions of years. When it was spotted in red giants in 1952 it proved that element-building was going on inside stars.

*temperature*   the degree of hotness of a body. Related to the energy of motion of the particles that compose it.

*triple-alpha process*   the process by which stars weld three helium nuclei into a nucleus of carbon, opening the way to the build-up of all heavy elements.

*weak nuclear force*   the second force experienced by protons and neutrons in an atomic nucleus, the other being the strong nuclear force. The weak nuclear force can convert a neutron into a proton, and so is involved in beta decay.

*x-process*   the mysterious process that the Burbidges, Fowler and Hoyle said had formed the light nuclei: deuterium, lithium, beryllium and boron.

*x-rays*   high-energy form of light.

# Select Bibliography

## Atoms

*Chemical Evolution*: Stephen Mason (Oxford, OUP 1992)
*A History of the Sciences*: Stephen Mason (New York, Collier 1962)
*Inward Bound*: Abraham Paris (Oxford, OUP 1994)
*Taming the Atom*: Hans Christian von Baeyer (Harmondsworth, Penguin 1994)
*The Discovery of Subatomic Particles*: Steven Weinberg (Harmondsworth, Penguin 1993)
*Thirty Years that Shook Physics*: George Gamow (New York, Dover 1985)
*One Two Three . . . Infinity*: George Gamow (New York, Dover 1988)
*The Great Physicists from Galileo to Einstein*: George Gamow (New York, Dover 1988)
*The Making of the Atomic Bomb*: Richard Rhodes (Harmondsworth, Penguin 1988)
*What Little I Remember*: Otto Frisch (Cambridge, Canto 1991)
*The Feynman Lectures in Physics*: ed. Robert Leighton et al. (Massachusetts, Addison-Wesley 1989)
*The World of Physics*: ed. Jefferson Hane Weaver (New York, Simon & Schuster 1987)
*The World Treasury of Physics, Astronomy and Mathematics*: ed. Timothy Ferris (Boston, Little, Brown 1989)
*Heisenberg's War*: Thomas Powers (Harmondsworth, Penguin 1994)
*Millikan's School*: Judith Goodstein (New York, W. W. Norton 1991)
*A Beginner's Guide to Hadronic Circuits*: Chris Illert (*Alchemy Today* vol. 2)

## Astronomy

*The Origin of the Chemical Elements*: R. J. Tayler and A. S. Everett (London, Wykeham 1975)

*The Stars: Their Structure and Evolution*: R. J. Tayler (London, Wykeham 1978)

*The Physical Universe*: Frank Shu (Mill Valley, University Science Books 1982)

*Home is Where the Wind Blows*: Fred Hoyle (Mill Valley, University Science Books 1994)

*Afterglow of Creation*: Marcus Chown (London, Arrow 1993)

*Exploring the Sun*: Karl Hufbauer (Baltimore, Johns Hopkins UP 1991)

*The Supernova Story*: Laurence Marschall (Princeton, Princeton UP 1994)

*The Nature of the Universe*: Fred Hoyle (Oxford, Blackwell 1950)

*100 Billion Suns*: Rudolf Kippenhahn (London, Counterpoint 1985)

*Cosmology*: Edward Harrison (Cambridge, CUP 1991)

*Supernovae and Nucleosynthesis*: David Arnett (Princeton, Princeton UP 1996)

*The Alchemy of the Heavens*: Ken Croswell (New York, Anchor 1995)

*The Anthropic Cosmological Principle*: John Barrow and Frank Tipler (Oxford, OUP 1990)

# Index